FM 31-70

US ARMY FIELD MANUAL

BASIC COLD WEATHER MANUAL

1959

CIVILIAN REFERENCE EDITION

UNABRIDGED HANDBOOK ON CLASSIC ICE AND SNOW CAMPING
AND CLOTHING, EQUIPMENT, SKIING, AND SNOWSHOEING FOR WINTER OUTDOORS

U.S. DEPARTMENT OF THE ARMY

Doublebit Press

New content, introduction, cover design, and annotations
Copyright © 2020 by Doublebit Press. All rights reserved.

Doublebit Press is an imprint of Eagle Nest Press
www.doublebitpress.com
Cherry, IL, USA

Original content under the public domain; unrestricted for civilian distribution. Originally published in 1959 by the U.S. Department of the Army.

This title, along with other Doublebit Press books are available at a volume discount for youth groups, clubs, or reading groups. Contact Doublebit Press at info@doublebitpress.com for more information.

Military Outdoors Skills Series: Volume 8

Doublebit Press Civilian Reference Edition ISBNs
Hardcover: 978-1-64389-158-3
Paperback: 978-1-64389-159-0

Doublebit Press, or its employees, authors, and other affiliates, assume no liability for any actions performed by readers or any damages that might be related to information contained in this book. Some of the material in this book may be outdated by modern standards. This text has been published for historical study and for personal literary enrichment. Remember to be safe with any activity that you do in the outdoors and to help do your part to preserve and be a good steward of our great American wild lands.

The Military Outdoors Skills Series
Historic Field Manuals and Military Guides on Outdoors Skills and Travel

Military manuals contain essential knowledge about outdoors life, thriving while in the field, and self-sufficiency. Unfortunately, many great military books, field manuals, and technical guides over the years have become less available and harder to find. These have either been rescinded by the armed forces or are otherwise out of print due to their age. This does not mean that these manuals are worthless or "out of date" – in fact, the opposite is true! It is true that the US Military frequently updates its manuals as its protocols frequently change based on the current times and combat situations that our armed services face. However, the knowledge about the outdoors over the entire history of military publications is timeless!

By publishing the **Military Outdoors Skills Series**, it is our goal at Doublebit Press to do what we can to preserve and share valuable military works that hold timeless knowledge about outdoors life, navigation, and survival. These books include official unrestricted texts such as army field manuals (the FM series), technical manuals (the TM series), and other military books from the Air Force, Navy, and texts from before 1900. Through remastered reprint editions of military handbooks and field manuals, outdoors enthusiasts, bushcrafters, hunters, scouts, campers, survivalists, nature lore experts, and military historians can preserve the time-tested skills and institutional knowledge that was learned through hard lessons and training by the U.S. Military and our expert soldiers.

Soldiers were the original campers and survivalists! Because of this, military field manuals about outdoors life contain essential knowledge about thriving in the wilds. This book is not just for soldiers!

This book is an important contribution to outdoors literature and has important historical and collector value toward preserving the American outdoors tradition. The knowledge it holds is an invaluable reference for practicing skills related to thriving in the outdoors. Its chapters thoroughly discuss some of the essential building blocks of outdoors knowledge that are fundamental but may have been forgotten as equipment gets fancier and technology gets smarter. In short, this book was chosen for Historic Edition printing because much of the basic skills and knowledge it contains could be forgotten or put to the wayside in trade for more modern conveniences and methods.

Although the editors at Doublebit Press are thrilled to have comfortable experiences in the woods and love our high-tech and light-weight equipment, we are also realizing that the basic skills taught by the old experts are more essential than ever as our culture becomes more and more hooked on digital technology. We don't want to risk forgetting the important steps, skills, or building blocks involved with thriving in the outdoors. This Civilian Reference Edition reprint represents a collection of military handbooks and field manuals that are essential contributions to the American outdoors tradition despite originating with the military. In the most basic sense, these books are the collection of experiences by the great experts of outdoors life: our countless expert soldiers who learned to thrive in the backwoods, deserts, extreme cold environments, and jungles of the world.

With technology playing a major role in everyday life, sometimes we need to take a step back in time to find those basic building blocks used for gaining mastery – the things that we have luckily not completely lost and has been recorded in books over the last two centuries. These skills aren't forgotten, they've just been shelved. *It's time to unshelve them once again and reclaim the lost knowledge of self-sufficiency.*

Based on this commitment to preserving our outdoors heritage, we have taken great pride in publishing this book as a complete original work. We hope it is worthy of both study and collection by outdoors folk in the modern era of outdoors and traditional skills life.

Unlike many other photocopy reproductions of classic books that are common on the market, this Historic Edition does not simply place poor photography of old texts on our pages and use error-prone optical scanning or computer-generated text. We want our work to speak for itself, and reflect the quality demanded by our customers who spend their hard-earned money. With this in mind, each Historic Edition book that has been chosen for publication is carefully remastered from original print books, *with the Doublebit Civilian Reference Edition printed and laid out in the exact way that it was presented at its original publication.* We provide a beautiful, memorable experience that is as true to the original text as best as possible, but with the aid of modern technology to make as beautiful a reading experience as possible for books that are typically over a century old. Military historians and outdoors enthusiasts alike are sure to appreciate the care to preserve this work!

Because of its age and because it is presented in its original form, the book may contain misspellings, inking errors, and other print blemishes that were common for the age. However, these are exactly the things that we feel give the book its character, which we preserved in this Historic Edition. During digitization, we ensured that each illustration in the text was clean and sharp with the least amount of loss from being copied and digitized as possible. Full-page plate illustrations are presented as they were found, often including the extra blank page that was often behind a plate. For the covers, we use the original cover design to give the book its original feel. We are sure you'll appreciate the fine touches and attention to detail that your Historic Edition has to offer.

For outdoors and military history enthusiasts who demand the best from their equipment, the Doublebit Press Civilian Reference Edition reprint of this military manual was made with you in mind. Both important and minor details have equally both been accounted for by our publishing staff, down to the cover, font, layout, and images. It is the goal of Doublebit Civilian Reference Edition series to preserve outdoors heritage, but also be cherished as collectible pieces, worthy of collection in any outdoorsperson's library and that can be passed to future generations.

*FM 31-70

FIELD MANUAL } HEADQUARTERS,
No. 31-70 } DEPARTMENT OF THE ARMY
WASHINGTON 25, D. C., *24 February 1959*

BASIC COLD WEATHER MANUAL

			Paragraphs	Page
CHAPTER	1.	INTRODUCTION	1, 2	3
	2.	INDIVIDUAL CLOTHING AND EQUIPMENT		
Section	I.	General	3–7	5
	II.	Clothing	8–17	11
	III.	Equipment	18–22	29
CHAPTER	3.	SMALL UNIT LIVING		
Section	I.	General	23, 24	35
	II.	Tentage and Other Equipment	25–34	35
	III.	Improvised Shelters	35–43	45
	IV.	Food and Water	44–57	56
	V.	Hygiene and First Aid	58–71	70
	VI.	Bivouac Routine	72–86	82
CHAPTER	4.	SKIING AND SNOWSHOEING		
Section	I.	Introduction	87, 88	98
	II.	Snow and Terrain	89–91	99
	III.	Military Skiing	92–121	102
	IV.	Military Snowshoeing	122–125	145
	V.	Application of Ski and Snowshoe Technique	126–130	149
CHAPTER	5.	MOVEMENT		
Section	I.	Problems Affecting Movement	131, 132	161
	II.	Foot Movement	133, 134	163
	III.	Trailbreaking	135, 136	164
	IV.	Land Navigation	137–139	175
	V.	Action When Lost	140–142	178
	VI.	Mechanized Aid to Movement	143–145	179
	VII.	Sleds	146, 147	186
	VIII.	Aircraft	148, 149	190

*This manual supersedes FM 31-70, 26 October 1951, including C 1, 15 December 1952, and FM 31-73, 30 September 1957.

			Paragraphs	Page
CHAPTER	6.	COMBAT TECHNIQUES		
Section	I.	The Individual and Northern Warfare	150, 151	192
	II.	Individual Weapons and Instruments	152–155	193
	III.	Fire and Movement	156–161	195
	IV.	Fighting Techniques	162–164	206
	V.	Camouflage and Concealment	165–175	214
	VI.	Mines and Obstacles	176–181	223
CHAPTER	7.	SMALL UNIT LEADERS		
Section	I.	General	182, 183	232
	II.	Peculiar Problems of Leaders	184, 185	234
APPENDIX	I.	REFERENCES		237
	II.	MAINTENANCE AND OPERATIONAL PROCEDURES FOR VEHICLES IN COLD WEATHER.		240
	III.	GROUND/AIR EMERGENCY CODE FOR USE IN AIR/LAND RESCUE SEARCH.		252
	IV.	SKI DRILL		257
	V.	EFFECTS OF COLD WEATHER ON INFANTRY WEAPONS.		269
	VI.	WEIGHTS OF COLD WEATHER CLOTHING AND EQUIPMENT.		277
GLOSSARY				282
INDEX				285

CHAPTER 1
INTRODUCTION

1. Purpose and Scope

a. This manual is designed to aid the individual soldier and small unit leader in fighting, living, and moving under the varying terrain and climatic conditions of the sparsely populated and relatively undeveloped cold areas of the world. As the majority of the sparsely populated and relatively undeveloped cold areas of the world are found in the Northern Latitudes, portions of this manual are specifically directed to those particular areas.

b. The material contained herein emphasizes that cold, with its attendant problems, affects military operations but does not prevent them. The proper use of authorized equipment and field expedients will, to a major degree, overcome any problems encountered as a result of cold. It is the combat commander's responsibility to train his men so they can make the environment *serve* military operations, *not hinder them*. This manual should be studied in connection with FM 31-71 and FM 31-72. The material presented herein is applicable to both atomic and nonatomic warfare.

2. Relation to Other Manuals

a. This manual supplements other manuals of the arms and services and is based on the assumption that normal individual and basic unit training have been completed.

b. FM 5-15 covers Field Fortifications.

c. FM 5-20 covers basic principles of camouflage.

d. FM 20-15 covers tents and tent pitching.

e. FM 21-10 covers military sanitation, including the adaptation of general procedures to the climatic conditions found in extremely cold areas.

f. FM 21-11 covers first aid, including the effects of cold and carbon monoxide poisoning.

g. FM 21-15 covers the care and use of individual clothing and equipment, including the packboard and head and mosquito nets.

h. FM 21-20 covers all aspects of physical training.

FM 21-76 includes information on health and first aid, orientation and traveling, water, foods, including plants and animals, firemaking and cooking, emergency living in cold weather areas, and similar information normally classified as bushcraft or outdoor living.

j. FM 22–100 covers command and leadership for the small unit leader.

k. FM 31–71 includes information of the northern environment and the effect the environment has on man and the works of man.

l. FM 31–72 includes information on mountain operations.

m. TM 5–295 covers military water supply.

n. TM 5–560 covers effects of cold or explosives.

o. TM 9–2855 is an instructional guide for the operation and maintenance of ordnance materiel in extreme cold.

p. TM 9–2300–203–12, Ordnance Maintenance, covers miscellaneous components for the Full Track Armored Infantry Vehicle M–59.

q. TM 10–228 covers fitting of footgear.

r. TM 10–275 covers the principles and utilization of cold weather clothing and sleeping equipment.

s. TM 10–530 covers principles of packing and rigging aerial delivery containers.

t. TM 10–703 and TM 10–708 cover individual and small detachment cooking.

u. TM 10–730 covers the operation and maintenance of Heater, Tent, Gasoline, 250,000 btu, Herman-Nelson (Model GT-3077) and Silent Glow.

v. TM 10–735 covers the operation and maintenance of Stove, Yukon, M1950.

CHAPTER 2
INDIVIDUAL CLOTHING AND EQUIPMENT

Section I. GENERAL

3. Basis of Issue

a. As used in this manual, individual clothing and equipment are those items issued or sold to a soldier for his personal use, and include certain organizational equipment utilized by the individual. The basis for issue of cold weather clothing and equipment may be found in TA 21, (Clothing and Equipment). Mandatory items of personal clothing are listed in AR 700-8400-1.

b. The US Army has the best clothing and equipment in the world. When properly fitted and properly utilized this clothing will provide adequate protection from the elements and will enable trained, well disciplined troops to carry out year-round field operations under cold weather conditions, wherever they may be encountered.

c. To utilize the protection afforded by the present standard cold weather clothing and equipment fully, it is necessary to understand the principles involved and the correct function of each item. This chapter covers basic principles and provides general guidance on the purpose and use of cold weather clothing and equipment. More detailed information is contained in publications listed in appendix I.

4. Individual Load

a. General.

   (1) This paragraph provides information on the various types of individual loads of clothing and equipment that must be worn, carried, and transported during winter operations in areas where the temperature ranges from moderate cold to extreme cold. Types of loads and their use are explained in *b* below. Weights of the various loads are shown in paragraph 5. A detailed breakdown of the weights of the individual items that comprise the various loads is contained in appendix VI.

   (2) Although it is highly desirable to move as much of the individual and unit field equipment as possible by unit

transportation, the individuals and the unit must be capable of carrying on cold weather operations without transportation. They must be able to live, fight, and move cross-country on foot, on skis, or on snowshoes, and to man-pack and man-haul (on 200-lb sleds) all of the equipment and supplies required to accomplish the mission and to provide protection from the cold.

b. *Types of Individual Loads.*
   (1) *Existence load.* The existence load consists of those items of clothing and equipment that must be worn and carried by the individual to enable him to live, move, and maintain himself (exist) in the field, for periods of short duration, under various cold weather conditions. The existence load is common to all soldiers regardless of their duty assignments.
   (2) *Battle load.* The battle load consists of that equipment which the individual uses for his contribution to the fighting capabilities of his small unit. It includes the individual weapon, ammunition, bayonet-knife, and hand grenades; it may include the protective mask and the steel helmet. The battle load is carried *at all times* by the combat soldier.
   (3) *Combat load.* The combat load consists of all the items in the existence load combined with all the items in the battle load. These items are required by the individual to enable him to perform his combat mission under various cold weather conditions.
   (4) *Protection and comfort load.* The protection and comfort load consists of those items of clothing and equipment required by the individual to sustain himself, increase his environmental protection, or increase his comfort for longer periods of time than provided by the existence load. It includes sleeping gear, toilet articles, towels, and additional items of clothing and equipment. These items are packed in both the rucksack and the duffle bag, the choice depending upon the weather, the mission, transportation available, distance to be covered, and many other factors. The rucksack may be man-packed by the individual in an emergency; however, every effort should be made to have these items carried on unit transportation. The additional items of clothing and equipment are required for both protection from the cold when a drop in temperature is expected and to protect the health of the individual by providing a change of clothing during prolonged periods in the field.

(5) *Full field load.* The full field load consists of all the items of clothing and equipment used by the individual during extended field operations under moderate to extreme cold weather conditions. The load includes all the items in the existence and battle (combat) load that are normally either worn or carried by the individual, plus the protection and comfort load that is usually carried on unit transportation. All of the items in the full field load are required when the individual or unit is engaged in cold weather field operations for extended periods.

5. Weights of Individual Loads

a. Figure 1 shows a breakdown of the types of individual loads of clothing and equipment required for field operations under various cold weather conditions, and give the weights of the various loads. Appendix VI gives weights of individual loads by item and points out those loads for which transportation should be provided whenever possible.

b. While the weights shown are exact, it is pointed out that the various load combinations are intended only as a guide, and it is emphasized that the decision as to what combinations of cold weather clothing and equipment best fit the needs of the unit for a particular mission rests squarely with the commander and small unit leader.

a—Combat load worn and carried in extremely cold weather.

Figure 1. The problem of the individual load.

Table of weights (a, fig. 1.)

```
EXISTENCE LOAD:
  Clothing _____ weight   23.97 lbs
  In pockets _____           7.22
  Skis and poles _____           9.50
                          Existence load total _____           40.69 lbs
BATTLE LOAD:
  Weapon, ammunition, grenades and entrench-
    ing tool _____          22.71
  Protective mask _____           3.30
                          Battle load total _____          26.01
COMBAT LOAD:            Combat load total _____          66.70 lbs
```

b—Full field load worn, carried, and transported in extremely cold weather. is expected to change from moderate cold to severe cold.

Figure 1—Continued.

Table of weights (b, fig. 1)

```
COMBAT LOAD: (same as for figure 1a.) _____ weight  _____   66.70 lbs
PROTECTIVE AND COMFORT LOAD:
  In rucksack (extra clothing for severe
    weather) _____  21.23
  In duffle bag (sleeping gear and extras) _____  30.92
                      Protective and comfort load total _____   52.15
FULL FIELD LOAD: Full field load total _____  118.85 lbs
```

c—Combat load worn and carried in extremely cold weather.

Figure 1—Continued.

Table of weights (c, fig. 1)

```
EXISTENCE LOAD:
    Clothing_____ weight   31.91 lbs
    In pockets (same as for figure 1a.)_____        7.22
    Skis and poles_____        9.50
                                Existence load total____           48.63 lbs
BATTLE LOAD: (same as for figure 1a.)_____     26.01
COMBAT LOAD:                   Combat load total_____           74.64 lbs
```

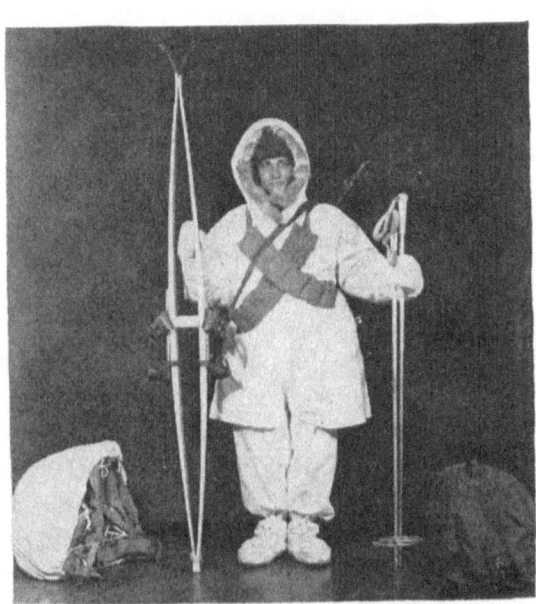

d—Full field load worn, carried, and transported in extremely cold weather.

Figure 1—Continued.

Table of weights (d, fig. 1)

```
COMBAT LOAD: (same as in figure 1c)_____ weight        74.64 lbs
PROTECTION AND COMFORT LOAD:
  In rucksack (sleeping gear)_____  26.86
  In duffle bag (extra clothing)_____  17.35
      Protective and comfort load total_____  44.21
FULL FIELD LOAD: Full field load total_____  118.85 lbs
```

6. Commander's Responsibilities

a. Many factors will influence the commander's decision as to what items of clothing and equipment his troops should wear or carry (fig. 1). These include the weather, mission at hand, actual duties to be performed, overall physical condition of the individuals and their degree of proficiency. If a movement is involved he must consider the distance to be traveled, the method of travel, and how the troops will be fed en route, if applicable. If the movement is on foot, he must bear in mind that under normal winter conditions, 65 to 75 pounds is the maximum weight a man can normally wear and carry and still be effective on reaching his destination.

b. The weight of the minimum individual clothing and equipment (to include skis, protective mask, water, and rations for one meal) required for *moderately cold weather* is approximately 45 pounds. The individual weapon, ammunition, and hand grenades increases the weight by another 15 to 20 pounds, depending upon the type of weapon and the amount of ammunition carried. For example: the total weight of an M-1 rifle, 2 bandoleers of ammunition, and two M26 fragmentation grenades is 18¼ pounds. Therefore, the individual combat load for field operations in moderately cold regions total approximately 65 pounds as compared to a combat load total of 40 to 45 pounds in temperate areas.

c. In addition to the individual combat load, another 45 to 55 pounds of clothing and equipment is required for the protection and comfort of each individual under conditions of *extreme cold*. Transportation must be provided for this additional load whenever possible.

d. The commander must take positive action to insure that a balance exists between what the individual is wearing and what he is required to carry in the way of equipment. He must also insure that troops dress as lightly as possible consistent with the weather in order to reduce the danger of excessive perspiring and subsequent chilling. The complete cold-wet or cold-dry uniform for the applicable environmental conditions must be readily available. A large proportion of cold weather casualties results from too few clothes being available to individuals when a severe change in the weather occurs.

7. Cold Weather Conditions

Two types of weather conditions affect the use of cold weather clothing: wet conditions and dry conditions.

a. Wet Conditions. Cold-wet conditions occur when temperatures are near freezing and variations in day and night temperatures cause alternate freezing and thawing. This freezing and thawing is often accompanied by rain and wet snow, causing the ground to become muddy and slushy. During these periods troops should wear clothing which consists of a water-repellent, wind-resistant outer layer and inner layers with sufficient insulation to provide ample protection in moderately cold weather (above 14° F.).

b. Dry Conditions. Cold-dry conditions occur when average temperatures are lower than 14° F. The ground is usually frozen and snow is usually dry, in the form of fine crystals. Strong winds cause low temperatures to seem colder and increase the need for protection of the entire body. During these periods troops should wear clothing which provides additional insulating layers.

Section II. CLOTHING

8. Purpose of Clothing

  a. Protection of Body Against Climatic Factors.
   (1) If the body is to operate efficiently, it must maintain a normal temperature. The body attempts to adjust itself to the variable external conditions it encounters. These attempts are evidenced by the need for more food to produce additional heat during colder weather, by perspiration to increase removal of heat during hot weather, and by the gradual darkening of the skin as protection against extended exposure to the rays of the sun.
   (2) Proper clothing, correctly worn, will assist the body in its adjustment to extreme climatic conditions. The clothing does this by holding the body heat in, thereby insulating the body against the cold outside air. Although the purpose of clothing is the same in all climates, the problem of protection becomes acute when freezing temperatures are involved. To understand this problem requires a knowledge of the methods by which the body resists the effects of climatic changes.

  b. Balancing Heat Production and Heat Loss. The body loses heat at variable rates. This heat may flow from the body at a rate equal to or greater than the rate at which it is produced. When heat loss exceeds heat production, the body uses up the heat stored in its tissues, causing a rapid drop in body temperature. Excessive heat loss is the cause of shivering. Although shivering uses body energy, it produces heat which, to a certain extent, counteracts the loss of energy and slows down the rate at which the body temperature will drop. The adding of proper clothing prevents excessive heat loss. To prevent overheating, clothing must be removed or adjusted.

   c. *Balancing Moisture.*

   (1) The body continuously passes a secretion and exhalation of fluid by the sweat glands of the skin. This secretion or exhalation of fluid is called perspiration. Excessive body heat and intense mental emotion cause perspiration to collect as moisture in the form of drops. When clothing prevents the evaporation of excessive moisture, it is always possible the moisture will freeze. In freezing temperatures it is as important to remove and adjust clothing to prevent excessive overheating as it is to add clothing to prevent heat loss.

   (2) The bare skin cools and freezes very rapidly when in contact with cold metal or liquids such as gasoline, fuel oils, or alcohol. In extreme cold or freezing temperatures accompanied by high winds, the skin can actually freeze before the deeper tissues cool. In the field a "buddy system" can be used where periodic checks are made of each other. Minor frostbites of the face are thawed by holding a warm bare hand against it. To thaw a frozen wrist, grasp it with the warm hand or put it inside the clothing and close to the body.

   d. *Protection of Body.* Clothing not only protects the body against cold but gives protection in various other ways. Outer clothing will repel, at least for a while, snow or rain. This outer clothing also reduces the effects of the wind. The springy quality of cold weather clothing gives excellent protection against the scratches of brush or tree branches and lessens the force of flying blocks and splinters of ice or frozen ground caused by projectiles.

9. Principles of Clothing Design

Certain principles are involved in the design of adequate cold weather clothing to control the loss of heat from the body, to facilitate moisture loss, and to protect the body.

   a. *Insulation.* Any material that resists the flow of heat is known as an insulating material. Dry air is an excellent insulator. Materials which hold quantities of motionless air are the best insulators. Woolen cloth contains thousands of tiny pockets within its fibers. These air pockets trap the air warmed by the body and hold it close to the skin. The principle of trapping air within the fibers of clothing provides the most efficient method of insulating the body against heat loss. Fur provides warmth in the same way; warm, still air is trapped in the hair and is kept close to the body.

   b. Layer Principles.
- (1) Several layers of medium-weight clothing provide more warmth than one heavy garment, even if the single heavy garment is as thick as the combined layers. The secret lies in the layers of air which are trapped between the layers of clothing. These air layers, as well as the minute air pockets within the fibers, are warmed by the body heat.
- (2) The layers of clothing are of different design. The wool underwear is most porous and has many air pockets. These air pockets trap and hold the air warmed by the body. To keep the cold outside air from reaching the still inside air that has been warmed by the body, the outer garments are made of windproof, water-repellent fabric.
- (3) The layer principle allows maximum freedom of action and permits rapid adjustment of clothing through a wide range of temperatures and activities. The addition or removal of layers of clothing allows the body to maintain proper body heat balance.

   c. Ventilation. Perspiration fills the air spaces of the clothing with moisture and reduces their insulating qualities. As perspiration evaporates, it cools the body just as water evaporating from a wet canteen cover cools the water in the canteen. To combat these effects, cold weather clothing is designed so that the neck, waist, hip, sleeve, and ankle fastenings can be opened or closed to provide ventilation. To control the amount of circulation, the body should be regarded as a house and the openings in the clothing as windows of the house. Cool air enters next to the body through the openings in the clothing just as cool air comes into a house when the windows are open. If the windows are opened at opposite ends of a room, cross-draft ventilation results. In the same way, if clothing is opened at the waist and neck, there is a circulation of fresh air. If this gives too much ventilation, only the neck of the garment should be opened to allow warm air to escape without permitting complete circulation.

10. Winter Use of Clothing

a. *Basic Principles of Keeping Warm.*
   (1) Keep clothing clean.
   (2) Avoid overheating.
   (3) Wear clothing loose and in layers.
   (4) Keep clothing dry.

b. *Application of Basic Principles.*
   (1) *Keep clothing clean.* This is always true from a standpoint of sanitation and comfort; in winter, in addition to these considerations, it is necessary for maximum warmth. If clothes are matted with dirt and grease, much of their insulation property is destroyed; the air pockets in the clothes are crushed or filled up and the heat can escape from the body more readily. Underwear requires the closest attention because it will be the dirtiest. Woolen underwear and socks should be washed in lukewarm water or, if warm water is not available, rinsed in cold water. Woolens should not be boiled or washed in hot water. When outer clothing gets dirty it should be washed with soap and water. All the soap must be rinsed out of the clothes, as soap left in the clothing will lessen the water-shedding quality of the clothing. In addition to destroying much of the normal insulation, grease will make the clothing more flammable.

   (2) *Avoid overheating.* In cold climates, overheating should be avoided whenever possible. Overheating causes perspiration which, in turn, causes clothing to become damp. This dampness will lessen the insulating quality of the clothing. In addition, as the perspiration evaporates it will cool the body even more. When indoors, a minimum of clothing should be worn and the shelter should not be overheated. Outdoors, if the temperature rises suddenly or if hard work is being performed, clothing should be adjusted accordingly. This can be done by ventilation (by partially opening parka or jacket) or by removing a layer of clothing. In cold temperatures it is better to be slightly chilly than to be excessively warm.

   (3) *Wear clothing loose and in layers.* Clothing and footgear that are too tight restrict blood circulation and invite cold injury. Wearing of more socks than is correct for the type of footgear being worn might cause the boot to fit too tightly. Similarly, a field jacket which fits snugly over a wool shirt would be too tight when a wool liner is

also worn under the garment. If the outer garment already fits tightly, putting additional layers under it will restrict circulation.

(4) *Keep clothing dry.*

  (a) Under winter conditions, moisture will soak into clothing from two directions—inside and outside. Dry snow and frost that collect on the uniform will be melted by the heat radiated by the body.

  (b) Outer clothing is water-repellent and will shed most of the water collected from melting snow and frost. The surest way to keep dry, however, is to prevent snow from collecting. Before entering heated shelters, snow should be whisked or shaken from uniforms and boots; it should not be rubbed off, as this will work it into the fabric. Whenever possible, footgear should be removed and the frost cleaned from the boots.

  (c) In spite of all precautions, there will be times when getting wet cannot be prevented and drying clothing may become a major problem. On the march, damp mittens and socks should be hung on the pack; even in below freezing temperatures, wind and sun will help dry this clothing. Damp socks or mittens may be placed under the parka, close to the body; the body heat will dry them out. In bivouac, damp clothing may be hung inside the tent, near the top, using drying lines or improvised drying racks. It may even be necessary to dry each item piece by piece by holding before an open fire. Clothing and shoes should not be dried too close to a fire; they might burn or scorch. Leather articles, especially boots, must be dried slowly. When drying mukluks, the insoles should be removed and dried separately.

11. Summer Use of Clothing

During the summer season the two main problems are external moisture (caused by rain, wet brush, muskeg) and insects. In the spring and autumn, protection against both cold and moisture is needed. During summer, spring, and autumn, the layer principles of clothing will remain the same, but the layers of cold weather clothing are reduced. Head net and leather gloves should be carried by the individuals at all times to provide necessary protection from swarms of insects.

12. Layer Components of Cold-Wet Weather Uniform (fig. 2)

a. Components. The basic components of the cold-wet weather uniform as illustrated in figure 2 are as follows:

(1) Undershirt, full length, sleeves natural.
(2) Drawers, ankle length, ribbed knit.
(3) Socks, wool, OD-9 w/cushion sole.
(4) Suspenders, trousers, scissors type, OG-107.
(5) Trousers, wool serge, OG-108, M-1951.
(6) Shirt, wool, OG-108.
(7) Trousers, cotton, wind-resistant, sateen, OG-107, water-repellent.
(8) Boot, combat, rubber, insulated, black, chevron cleated sole and heal (or Boot, combat, leather, russet).
(9) Coat, cotton, sateen, wind-resistant, water-repellent-treated, OG-107, slide fastener closure.
(10) Liner, coat, natural, mohair, frieze.
(11) Cap, field, cotton poplin, wind-resistant, olive green.
(12) Glove shells, leather, M-1949 (or Glove shells, leather, black w/wrist strap and buckle with Glove inserts, wool, olive drab, shade #30A, M-1949).
(13) Hood, winter, olive green, cotton, shade 107.
(14) Poncho, coated nylon, OG shade 107 (not illustrated).

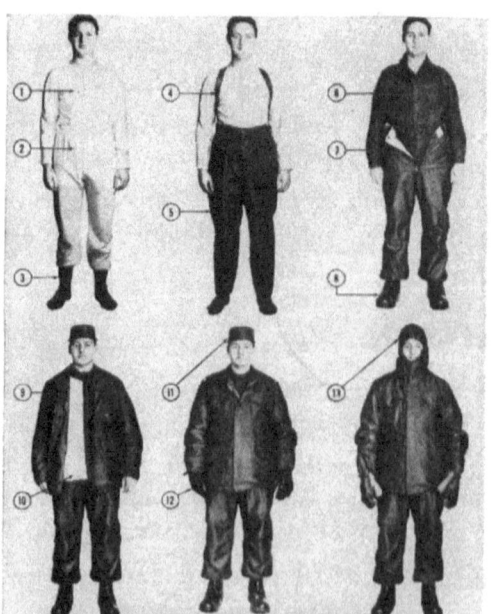

Figure 2. Basic components of cold-wet weather uniform.

b. Underwear. The same underwear is worn for either cold-dry or cold-wet conditions. It is loose-fitting and made of cotton, wool, or a combination of the two. The drawers, wool trousers, and/or trousers, cotton, wind-resistant, are worn loosely at the waist, supported by suspenders from the shoulders, so that neither circulation nor ventilation is restricted.

c. Intermediate Layer. The wool trousers and shirt provide excellent insulation against cold. The shirt may be worn either tucked inside the trousers for additional warmth or outside the trousers for better ventilation. The trousers are not designed to be worn as an outer garment under field conditions since they lose their insulating qualities if they become wet or matted with dirt.

d. Outer Layer.
    (1) *Coat.* The coat or jacket is made up of a shell and a detachable liner. The liner has a smooth surface which is worn facing the body and which reduces drag and binding. The shell and liner are worn directly over the intermediate layer in cold weather.
    (2) *Trousers.* The trousers are made of smooth, light, water-repellent and wind-resistant cotton material. They have extra closures and adjustments to provide for ventilation and better fit.

e. Headgear.
    (1) *Cap.* A lightweight cotton cap or a heavier pile cap is issued to protect the head. Each cap may be worn alone or under the hood. Each cap has a visor and earflaps. The visor must be raised when using sighting equipment.
    (2) *Hood.* A lightweight hood is issued and worn for protection against the wind and snow. The hood is not worn over the steel helmet. Instead, the helmet is placed over the hood by adjusting the head band on the helmet liner. In extreme low temperatures the use of helmets will normally be restricted to static situations only.

f. Handgear (fig. 3).
    (1) *Gloves.* Standard black leather gloves are worn in mild weather, or when work must be done that requires more freedom of finger movement than can be acquired with heavier handgear. In colder weather the same gloves are worn with wool inserts (fig. 3). Special cotton insert gloves are worn in summer months for protection against insect bites.
    (2) *Mittens.* The three-finger mittens (fig. 3) are worn in cold weather with wool trigger-finger inserts. The shells alone may be worn in mild weather.

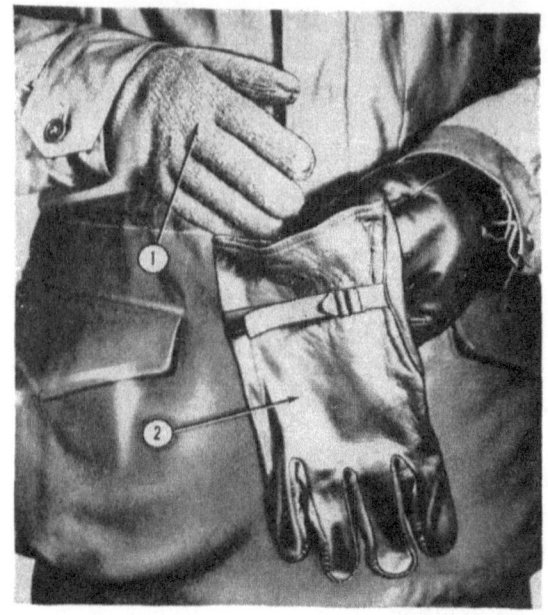

1 Wool insert   2 Leather shell

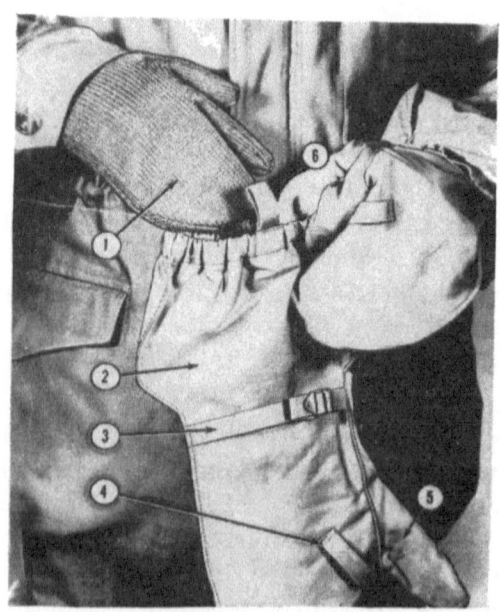

1 Mitten insert   3 Adjustable wrist strap   5 Trigger finger
2 Mitten shell   4 Trigger finger loop   6 Cord loop

Figure 3. Types of gloves and mittens.

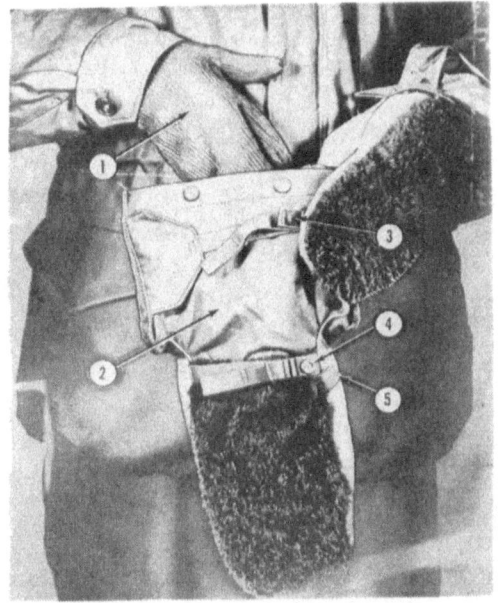

1 Mitten insert
2 Mitten shell
3 Adjustable gauntlet strap
4 Snap fastener
5 Cord loop

Figure 3—Continued.

13. Layer Components of Cold-Dry Weather Uniform
(fig. 4)

a. Components. The basic components of the cold-dry weather uniform as illustrated in figure 4 are as follows:

- (1) Undershirt, full length, sleeves natural.
- (2) Drawers, ankle length, ribbed knit.
- (3) Socks, wool, OD-9, w/cushion sole.
- (4) Socks, wool, natural.
- (5) Socks, wool, felt, white.
- (6) Suspenders, trousers, scissors type.
- (7) Shirt, wool, OG-108.
- (8) Liner, trousers, field.
- (9) Trousers, cotton, wind-resistant, sateen, OG-107, water-repellent.
- (10) Boot, mukluk, cotton duck, rubber sole, natural (or Boot, combat, rubber, insulated, white (not illustrated)).
- (11) Cap, field, cotton poplin, wool pile lining, OG-107.
- (12) Liner, coat, natural, mohair, frieze, OG-107.
- (13) Coat, cotton, sateen, wind-resistant, water-repellent-treated, OG-107, slide fastener closure.

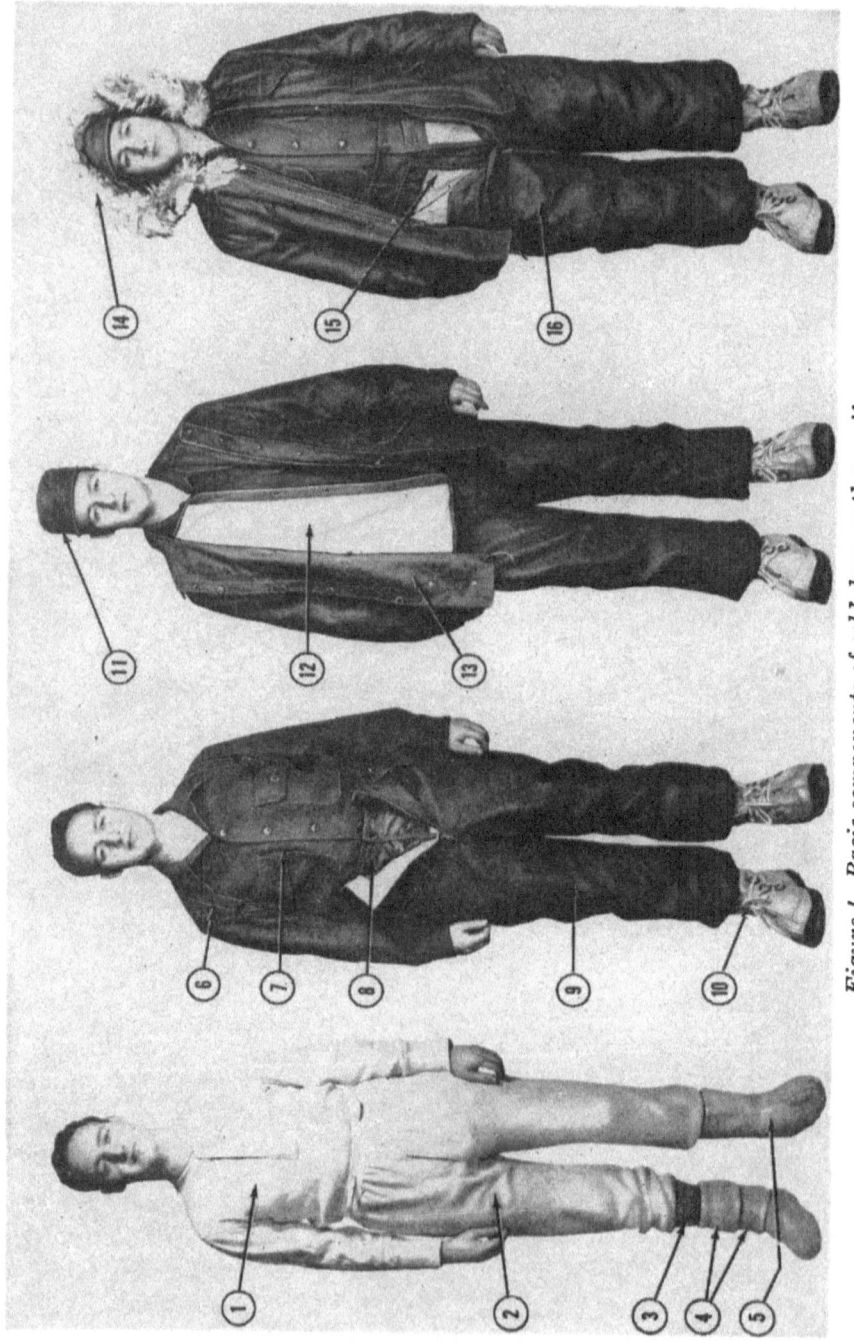

Figure 4. Basic components of cold-dry weather uniform.

20 AGO 10048B

(14) Hood, winter, olive green.
(15) Liner, trousers, arctic, mohair, frieze, natural.
(16) Trousers, cotton, OG–107, M–1951.

b. Underwear. The same underwear is worn for cold-dry conditions as for cold-wet weather (par. 10*b*).

c. Intermediate Layer. The wool shirt is worn with the cold-dry uniform. The wool trousers are replaced by a wool liner which is worn inside the trouser shell, smooth side next to the underwear. The coat with detachable wool liner, used as an outer layer in the cold-wet uniform, is worn as an intermediate layer in cold-dry conditions.

d. Outer Layer.
  (1) *Arctic trouser and liner.* An additional pair of shell trousers, known as arctic trousers, and a wool liner which is attached inside the shell, are worn over the intermediate trouser and liner when the soldier must remain inactive (in a listening post, for example) and when temperatures of —50° F. to —60° F. are encountered.
  (2) *Parka.* The parka is made in much the same way as the coat and has a similar detachable wool liner with one smooth side. This smooth surface allows it to be slipped over or off the intermediate layers with very little difficulty. The parka is longer and roomier than the coat and has a number of adjustable features. It is always worn as an outer garment and may be worn over the coat and liner.

e. Headgear.
  (1) *Caps.* The same caps may be worn under cold-dry conditions as are worn for the cold-wet climate. Choice in selection of the correct type of headgear consistent with the temperature conditions will also preclude certain cases of overheating. As a general rule, in temperatures down to zero degrees with a negligible amount of wind, the cotton field cap with ear flaps is best for most individuals. For weather colder than zero degrees, the cap with wool pile lining provides the best protection against the elements. Use of the pile field cap in warmer weather will increase the chances of an individual becoming overheated.
  (2) *Hood.* The arctic hood is one of the most valuable items of the cold-dry uniform and may be worn attached to either the coat or the parka. It is adjusted for use, as shown in 1, figure 5, for maximum protection against cold wind.

A malleable wire inside the fur ruff may be shaped as desired for visibility or greater protection of the head and face. The hood may be turned back as shown in the same figure. An elastic band on top of the hood may be placed over the forehead and adjusted to keep the hood in a position of maximum protection. It may be necessary to beat frost from the fur ruff from time to time. The steel helmet is not normally used with the arctic hood. Fastening of hood to the parka is illustrated in 1, figure 5.

(3) *Hood discipline.* Unit commanders must enforce "hood" discipline, especially while men are on sentry duty or on patrols. The arctic hood and the pile cap with flaps down will greatly reduce a man's hearing capabilities. When the temperature or wind does not require the use of heavier headgear, the lightweight gear should be worn. The hood should be removed before the head starts to perspire. Breathing into the hood causes moisture and frost to accumulate and should be avoided as much as possible. Accumulated frost should be removed frequently.

f. *Handgear.*

(1) All gloves previously discussed may also be worn with the cold-dry uniform when temperatures are appropriate.

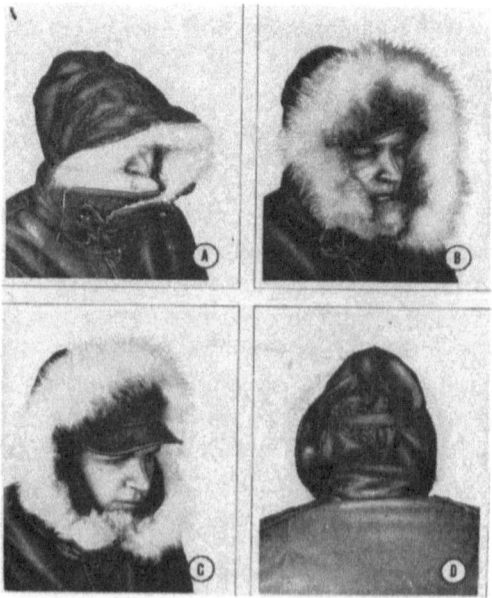

1—*Adjustment.*

Figure 5. Adjusting and fastening hood, winter.

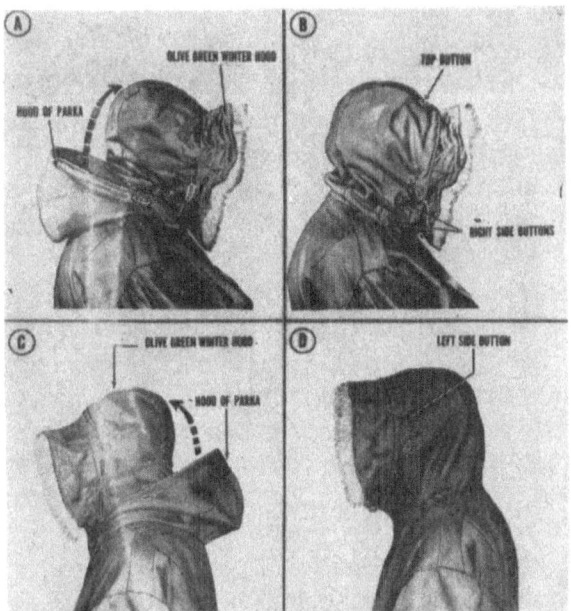

2—*Fastening.*
Figure 5—Continued.

(2) For extreme cold, arctic mittens are worn (fig. 3). The mitten has a pile back which may be used to warm the face. A detachable wool liner is snapped inside the shell. The mitten has several adjustable features and is worn with a cord which passes over the shoulders so that the mittens will not be lost. They also incorporate a snap fastener so that the mittens may be snapped together and worn behind the back when not in use. In this manner they are kept out of the wearer's way, yet are still available for immediate use. The trigger-finger wool insert may also be worn inside the mitten and liner for additional protection against cold.

(3) Personnel engaged in delicate finger operations, such as instrument adjustment, may be issued lightweight anti-contact gloves. These gloves prevent the skin from sticking to cold metal and allow for finger dexterity. They will provide protection against cold for only a very short period.

(4) The general rules concerning the use of clothing apply also to the handgear—keep it clean, avoid overheating, wear loose in layers, and keep it dry.

  (a) The outer shell should always be worn with the minimum insulation necessary to provide protection, thus avoiding perspiration. Inserts should never be worn by themselves, as they wear out quickly. Trigger-finger inserts are designed to fit either hand. Changing them to opposite hands frequently will insure even wear.

(b) Tight-fitting sleeves should be avoided. They may cut down circulation and cause hands to become cold.

(c) When handling cold metals, the hands should be covered to prevent metal burns.

(d) To keep hands warm when wearing mittens, the fingers should be curled (inside the mittens) against the palm of the hand, thumbs under the fingers, or flexed inside the mitten whenever possible to increase the blood circulation. Hands may be exercised by swinging the arms in a vertical circle. Frostbitten hands can be warmed by placing them next to the skin under the armpits.

(e) An extra pair of inserts should be carried.

(f) The mitten shells should be used without the inserts if the weather is not too cold. In colder weather the wool trigger-finger mitten insert must be worn under the trigger-finger mitten shell. For extreme cold weather the Mitten Set, Arctic, is used.

(g) Cold weather clothing will be utilized in accordance with the principles outlined in TM 10-275.

14. Footgear

a. General. The feet are more vulnerable to cold than are other parts of the body. Cold attacks feet most often because they get wet easily (both externally and from perspiration) and because circulation is easily restricted. Footgear is the most important part of cold weather clothing.

b. Principles.

(1) The rule of wearing clothing loose and in layers also applies to footgear. The layers are made up by the boot itself and by different combinations of socks and insoles. Socks are worn in graduated sizes. The instructions pertaining to fitting of footgear, as outlined in TM 10-228, must be carefully adhered to. If blood circulation is restricted, the feet will be cold. Too many socks, worn too tightly, might easily mean freezing of the feet. For the same reason: AVOID LACING FOOTGEAR TIGHTLY.

(2) The rules about avoiding overheating and keeping dry are difficult to follow, so far as feet are concerned, because the feet perspire more readily than any other part of the

body. Footgear is more often subjected to wetting than are other items of equipment. Shoes are designed to decrease this disadvantage as much as possible. A dry change of socks should be carried at all times. Whenever the feet get wet, socks should be changed as soon as possible. If wearing mukluks, both socks and insoles should be changed when they become wet. Footgear should be dried at the first opportunity.

(3) Footgear should be kept clean. Socks should be changed when they become dirty. Socks and feet should be washed frequently. This washing will help keep feet and socks in good condition.

(4) The feet should be exercised. Stamping the feet, double-timing a few steps back and forth, and flexing and wriggling toes inside the boots will require muscular action, which produces heat, and will help keep the feet warm. The feet should be massaged when changing socks.

(5) Boots are designed to be attached to individual oversnow equipment (skis and snowshoes). SKI BINDINGS OR OTHER BINDINGS MUST BE ADJUSTED CAREFULLY. If they are too tight, the circulation of blood is restricted and feet will get cold. Improperly adjusting bindings may soon chafe feet or badly wear and tear the boot.

c. *Types.*

(1) *Boots, Combat, Rubber, Insulated, (Black).* These boots (1, fig. 6) are particularly useful in snow, slush, mud, and water (cold-wet conditions), but are not adequate for prolonged wear in temperatures below —20° F. They are specifically designed for combat personnel who may not have the opportunity to frequently change to dry socks. Insulating material is hermetically sealed into the sides and bottoms of the boots. The insulation takes the place of removable insoles and the secondary layer of socks worn in other types of cold weather boots. Perspiration from the feet and water spilling over the tops of the boots cannot reach the insulating material because it is sealed-in and always remains dry. Moisture from outside sources or from perspiration may make the socks damp; this dampness is not harmful to the feet, provided they receive proper care such as frequent drying and massaging. Only one pair of cushion-sole socks should be worn with the boots. Additional socks should not be worn as the feet may become cramped, resulting in restricted

blood circulation and cold feet. The type of socks to be worn by the individual should be taken into consideration in fitting the boot.

(2) *Mukluks.* Mukluks (2, fig. 6) are the warmest type of of footgear issued for use under cold-dry conditions. They should be worn in temperatures below —20° F. Frost does not readily collect on mukluks because the material is not waterproof. For this reason, these boots should not be worn in temperatures above 10° F. They should be used in deep, dry snow, but never in wet snow. The following sock combination should be worn with the type mukluks illustrated: one pair of cushion-sole socks, two pair of woolen ski socks, one pair of white felt socks, and two pair of felt insoles. The mukluk has no arch support and will cause the wearer's feet and legs to tire very easily. They are less suitable for use in spring than other types of footgear.

(3) *Boot, combat, rubber insulated (white).* The insulated white rubber boots have been designed to replace the mukluk. They are particularly useful under cold-dry conditions and will protect feet in temperatures as low —60° F. They are specifically designed for personnel who, at the present time, are frequently required to change to dry socks. The boots are similar in design to the black rubber insulated boot, except that the white boots are for use in extremely cold weather. The wool insulating

1—*Boot, Combat, Rubber, Insulated (Black).*

Figure 6. Footgear.

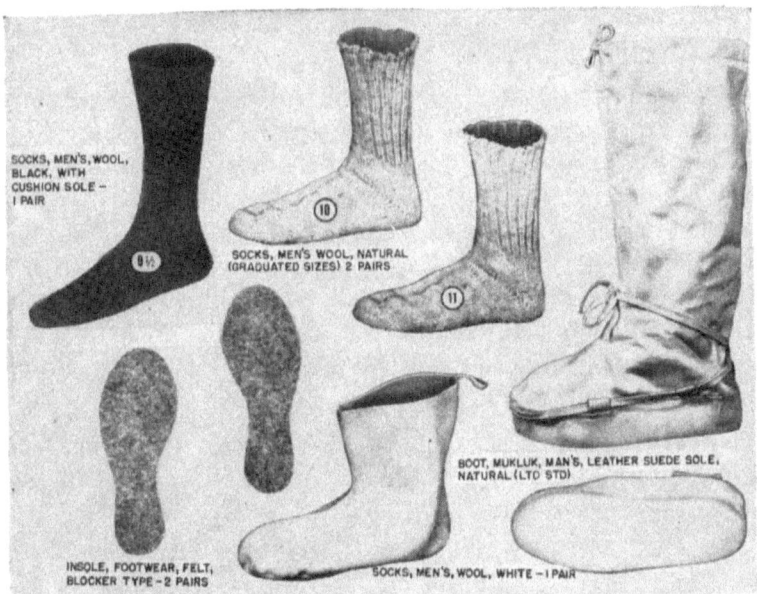

2—*Boot, Mukluk, Leather Suede Sole with Socks and Innersoles.*

Figure 6—Continued.

material is sealed between two waterproof layers of rubber. For this reason, perspiration from the feet or moisture on the outside cannot get into the insulation to destroy its insulating properties. The moisture from outside and perspiration from the feet may make the socks damp, but the dampness is not harmful to the feet as long as they receive proper drying and massaging whenever possible. When possible, socks should be changed daily. Only one pair of wool cushion-sole socks is worn with the boots. The boots should be inspected daily for holes or punctures.

15. Nose and Cheek Protectors and Masks

a. A nose and cheek protector may be issued for use during severe windchill conditions. Protectors should be removed at intervals to check for frostbite.

b. A certain amount of protection can be gained by covering as much of the face as possible with the wool scarf. It may be adjusted from time to time, and should be rotated when the section opposite the mouth and nose becomes covered with frost. The frozen end should be left outside the parka.

16. Camouflage Clothing

a. Winter camouflage clothing consists of white trousers and lightweight parka with hood. In addition, white covers for mittens and rucksacks may be issued.

b. Camouflage clothing provides a means of concealment and camouflage from the enemy—both from the ground and from the air—in winter conditions. Use of the white camouflage clothing is, however, dependent on the background; generally speaking, on vegetation and the amount of snow on the ground. The complete white suit (fig. 128) is worn when terrain is covered with snow, consideration being given to both ground and air observation. Mixed clothing (fig. 129)—white parka and dark trousers, or vice versa—is used against mottled backgrounds. The correct use of camouflage clothing is extremely important.

c. Camouflage clothing may become frosty and icy after use. As with all clothing, the frost and ice must be removed to expedite drying. Soiled camouflage clothing will lose its effectiveness; therefore, care must be exercised when handling sooty stoves, digging ground, and performing similar tasks. Avoid scorching or burning the garments when drying or when lying down by an open fire. The clothing should be washed or changed frequently. When changing, clothing should be checked to insure that it fits over the basic garments without restricting movement.

17. Maintenance of Clothing and Equipment

Clothing and equipment must be kept clean. Dirt and grease will clog the air spaces of clothing and reduce its insulation properties. If available, cotton underwear may be worn next to the body to keep the woolens clean longer. Woolen socks and underwear should be washed in lukewarm water and stretched back to size before drying. Cold water may be used if necessary. Soap will reduce the insulating qualities and should be thoroughly washed out. Rips and tears in clothing should be mended as soon as possible. Safety pins or tape may be used as a field expedient for this purpose until more permanent repairs can be effected. Mildew may be prevented by frequent drying.

 a. Footgear.
  (1) *Boots.* The leather in boots should be treated with approved agents. Mukluks should be spot-cleaned with a damp cloth and soap. Normally, the rubber combat boot can be repaired with ordinary tire patching or air mattress patching material. If these items are not readily available, friction tape or even chewing gum may be used temporarily to plug up the hole and prevent moisture from damaging the insulation. If the damage cannot be repaired, the boots should be removed, air-dried, and

turned in for replacement as soon as possible. The inside of the boots should be washed at least once a month.

    (2) *Socks.* Socks should be washed daily, using lukewarm water to avoid excessive shrinkage. After washing, they should be wrung out and stretched to natural shape before drying. Holes in socks should be repaired as soon as possible, taking special precautions to avoid bunching or roughness of the mended area. Improperly mended socks may cause blistered feet.

  b. Handgear. Holes should be mended promptly. Gloves or mittens should not be dried too near an open fire.

  c. Headgear. Headgear should be washed as required to remove perspiration, dirt, and hair oils. When drying, normal care must be exercised to avoid scorching or burning.

Section III. EQUIPMENT

18. Sleeping Equipment

  a. The complete sleeping bag for use in cold climates consists of three parts: a case, of water-repellent material; an inner bag (mountain type), of quilted construction, filled with a mixture of down and feathers; and an outer bag (Arctic bag), of the same material as the inner bag. In addition, an insulating air mattress and a waterproof bag into which the sleeping bag is packed may be issued.

  b. When temperatures are normally above 14° F., only one bag is used. It is placed in and laced to the cover. When temperatures are below 14° F. both bags are used. The inner is placed in the outer bag and the outer bag laced to the cover.

  c. When the bag is used, it is first fluffed up so that the down and feather insulation is evenly distributed in channels, thus preventing matting. Since cold penetrates from below, and the insulation inherent in the bag is compressed by the weight of the body, additional insulation is placed under the bag whenever possible. If an insulated sleeping pad is not available, added insulation can be obtained by placing dry shelter halves (ponchos), extra clothing, packboards, fiber ammunition or food containers, or boughs between the sleeping bag and the ground. The insertion of a waterproof cover, such as a poncho, between the sleeping bag and the sleeping pad will prevent the pad and bag from freezing together at very cold temperatures. Care must be taken to prevent puncturing sleeping pads or damaging sleeping bags. In general, the more insulation between the sleeping bag and the ground, the warmer the body.

d. If the tactical situation permits, individuals should avoid wearing too many clothes in the sleeping bag. When too many clothes are worn in the sleeping bag they tend to bunch up, especially at the shoulders, thereby restricting circulation and inducing cold. Too many clothes also increase the bulk and place tension upon the bag, thus decreasing the size of the insulating air spaces between layers and reducing the efficiency of the insulation. In addition, too many clothes may cause the soldier to perspire and result in excessive moisture accumulating in the bag, a condition which will likewise reduce the bag's insulating qualities. Extra clothing should be placed under the bag.

e. The sleeping bag should be kept dry. It should be opened wide and ventilated after use to dry out the moisture that accumulates from the body. Wherever possible, it should be sunned or aired in the open. The bag should always be laced in its water-repellent case and carried in the waterproof bag to prevent snow from getting on it. The warmth of the body could melt the snow during the night and cause extreme discomfort. Individuals should avoid breathing into the bag. If the face becomes too cold it should be covered with an item of clothing.

f. The sleeping bag is a useful and comfortable item, but it is possible to live without it. During long cross-country movements, carrying heavy loads of extra ammunition or rations, it may be necessary to leave it behind temporarily. Therefore it may be necessary to sleep without the bag, in a heated tent or by the open fire, even in extreme cold temperature.

g. The bag is equipped with a quick-release slide fastener. Under ordinary conditions the bag is closed and opened by pulling the slide. In an emergency, however, the slide can be jerked quickly over the upper end, releasing the entire closure at once.

19. Manpack Equipment

a. Rucksack. The rucksack (1, fig. 7) is the normal pack equipment used by troops in cold areas. It consists of a canvas bag with a map pocket in the flap, a metal tubing pack frame, a webbing backstrap, a belly strap, and a white camouflage cover. It can be used with or without the pack frame. Medium weight loads (30 to 50 pounds) can be carried more easily in the rucksack than in an ordinary pack. Most of the weight is distributed around the hips so the shoulders take only part of the weight. This gives a low center of gravity and allows the free upper-body movement which is necessary for skiing. The frame holds the load off the wearer's back, allowing air to circulate between the clothes and the load, thus reducing the amount of perspiration that collects on the back.

The belly strap prevents the load from swinging to the sides in skiing turns.

b. Packboard. Packboards (2, fig. 7) are generally constructed of plywood. A canvas back-section acts as a cushion against the body and affords needed ventilation. Metal hooks are fastened to either side of the packboard to provide anchorage for lashing. Attachments and quick-release straps are also issued with packboards to make them suitable for transport of heavy weapons and ammunition. Packboards can be used to carry loads from 50 to 100 pounds. They are better suited for troops moving on foot or on snowshoes than for troops moving on skis, since the balance and body movement so necessary for skiing are somewhat impaired by the high center of gravity the packboard brings about.

c. Pistol Belts. Ammunition should be carried in bandoleers when possible. However, when the pistol belts are worn for this purpose they should be worn loosely. Items such as compasses and first aid packets may be carried in pockets or attached by means of the eyelets to the rucksack or cartridge belt.

20. Miscellaneous Equipment

a. Sunglasses. Sunglasses should always be worn on bright days when the ground is covered with snow. They are designed to protect the eyes against sun glare and blowing snow. If not used,

1—*Rucksack.*

Figure 7. Rucksack and packboard.

2—*Packboard.*

Figure 7—Continued.

snow blindness may result. They should be used when the sun is shining through fog or clouds. A bright, cloudy day is deceptive and can be as dangerous to the eyes as a day of brilliant sunshine. The sunglasses should be worn to shade the eyes from the rays of the sun that are reflected by the snow. When not being used, they should be carried in the protective case to avoid scratching the lens or breaking. If sunglasses are lost or broken, a substitute can be improvised by cutting thin, inch-long slits through a scrap of wood or cardboard approximately 6 inches long and 1 inch wide. The improvised sunglasses (fig. 9) can be held on the face with strips of cloth if a cord is not available.

b. Canteens and Thermos Bottles. The conventional canteen is carried in a fabric bag; however, this will not keep the liquid in the canteen from freezing in extreme cold. When possible, the canteen should be carried in one of the pockets or wrapped in any woolen garment and packed in the rucksack next to the back of the wearer. If available, warm or hot water should be placed in the canteen before starting on a march. During extreme cold the canteen should never be filled over $2/3$ full. This will allow room for expansion, if ice should form, and will prevent the canteen from rupturing. Insure that the cork gasket is in the cap of the canteen at all times. This is an important precaution and will prevent the liquid from leaking out and dampening the clothing in the rucksack. Conventional thermos bottles will keep the liquid hot, or at

Figure 8. Improvised sunglasses.

least unfrozen, over a day's march to the next camp. If canteens or thermos bottles freeze, they should be thawed out carefully to prevent bursting. The top should be opened and the contents allowed to melt slowly.

   c. Pocket Equipment. There are several small items that should be carried in the pockets so they will be readily available for use. Having these items when they are needed will contribute to the general well-being of individuals and help prevent injuries. A good sharp knife is an essential item. It is useful for cutting branches, in shelter construction, in repairing ski bindings, and numerous other tasks. The waterproof matches should be kept in the watertight matchbox and used only in an emergency. They should never be used when ordinary matches and lighters will function. Sunburn preventive cream will protect the skin from bright, direct sunshine, from sun rays reflected by the snow, and from strong winds usually encountered at high altitudes. The chapstick will prevent lips from chapping or breaking due to cold weather or strong winds. Insect repellent protects the exposed skin areas from insect bites. The emergency thong has numerous uses, such as lashing packs, replacing broken shoelaces, and repairing ski and snowshoe bindings. Fire starters, M1, are used to start wood fires when other fire starting means fail, and are extremely useful when only wet or frozen wood is available.

21. Steel Helmet

   The steel helmet may be worn during warm periods in cold areas in the same manner as in more moderate climates. It may also be worn with the lightweight hood during moderately cold weather by adjusting the headband on the helmet liner and placing it over the lightweight hood. During extreme low temperatures the use of the steel helmet will normally be restricted to short periods during static situations only.

22. Protective Mask

a. The standard issue protective mask may be worn in warm to moderately cold weather in the same manner as in more moderate climates. When the mask is used in extreme cold, the rubber facepiece should be warm enough to make it pliable when it is adjusted to the wearer's face. One method of keeping the mask warm is to carry it inside the outer garments and next to the body. It is also recommended that the mask be kept inside the sleeping bag during the night. Another precaution that may reduce the chances of frostbite is to place pieces of tape over the metal rivet heads where they touch the face. When this is done, special care must be exercised to avoid air leaks when adjusting the mask. On removing the mask, any moisture on the face should be wiped off immediately to prevent frostbite. After drying the face, the facepiece of the mask should be thoroughly dried to prevent freezing of moisture inside the mask. The rubber cover of the outlet valve should also be raised and the valve, surrounding area, and the inside of the cover wiped dry to prevent the outlet valve from icing.

b. If it becomes necessary to wear the mask for protection against chemical agents during extreme cold weather, troops must be advised that the facepiece of the protective mask will not protect the face from the cold and that, in fact, the opposite is true. The danger of frostbite increases when the mask is worn.

c. The Army has successfully tested and standardized a winterization kit (Winterization Kit, Protective Mask, M1) which will make the M9A1 or the M9 mask wearable in extreme cold at temperatures down to —40° F. Details of the kit and its use are outlined in TM 3–522–15 and TB 3–205–2. The kit consists of a hood, insulating eyepiece lenses, antisnow glare lenses, a cheek pad, and spare nose cup valves. The kit provides extra insulation of the mask and the dehumidified exhaled breath is vented from the bottom of the hood thereby warming the head and preventing CBR agents from leaking into the hood.

CHAPTER 3
SMALL UNIT LIVING

Section I. GENERAL

23. Characteristics of Operations in Cold Weather

Unlimited space and a sparse, widely scattered population are dominant features of most of the colder regions of the world. Such conditions permit unrestricted maneuver for troops properly trained and equipped for cold weather operations. Warfare under such circumstances is characterized by small, widely dispersed forces operating at great distances from other small units or their parent organization. Units must be highly mobile and have the ability to sustain themselves while carrying out independent operations over extended periods of time.

24. Composition of Units

a. Small units (squad, gun crew, tank crew, wire team, etc.) form the basic working group for cold weather operations. Under normal operating conditions they will work together, cook and eat together, and share the same tent or other shelter. These small units should be formed at the beginning of training and, if possible, kept intact. The standard to be achieved is a unit which can make or break camp quickly, efficiently, and silently under all conditions; one in which each man knows the tasks to be completed and does them without having to be told.

b. Small units operating in cold weather must be thoroughly familiar with the special equipment required and the techniques involved in living away from their parent organization for extended periods of time. Equipment, and the techniques of using it, are discussed in this chapter.

Section II. TENTAGE AND OTHER EQUIPMENT

25. General

A considerable quantity of various types of special equipment is required to maintain small units in cold weather. Permanent shelters are usually scarce in the areas of operations and heated tentage is required, regardless of the season of the year. Special type tools are necessary for establishing bivouacs, breaking trails, and constructing temporary winter roads and battle positions.

26. Need for Shelter

a. In order to conduct successful military operations in cold weather and maintain a high level of combat efficiency and morale, heated shelter must be provided for all troops. An individual's ability to continue to work, live, move, and fight under extreme climatic conditions depends upon adequate shelter. Tents and stoves, therefore, become a vital part of cold weather equipment.

b. In summer the requirement for shelter is still present and must not be overlooked. The many insects, cool nights, wind, and the necessity for drying wet clothing and equipment, emphasize this requirement.

c. In cold weather, tents should be placed as close as practicable to the scene of activity, whether the activity be combat or administrative. By so placing the tents, rotation of men for warmup is possible and maximum continuity of effort can be maintained.

d. Tents vary in size and shape, depending on their purpose. Small units such as a rifle squad, artillery section, or similar type unit are normally equipped with one 10-man arctic tent. During combat, fewer tents will be needed, as part of the personnel are always on guard detail, occupying positions, or performing similar missions. It may become necessary for the unit, temporarily, to use only one-half or one-fourth of its tentage; i.e., one 10-man tent per platoon, with the men sleeping on a rotation basis. Reduced numbers of tents and stoves will decrease the requirement for logistical support, such as fuel and transportation. Elements smaller than the rifle squad or artillery section, which require less shelter space, are normally equipped with the 5-man tent.

e. Normally, small reconnaissance patrols are not equipped with tents, as tents tend to hamper the mobility and speed of the patrol. Strong combat patrols and long-range reconnaissance patrols may be equipped with tents and stoves if sufficient transportation is available to move the extra weight. When speed is of the essence, patrols will improvise shelters built from local materials at hand. For semipermanent base camps, portable type frame shelters may be erected for increased comfort of the troops.

27. Description of Tentage

a. General. Tentage issued for use in cold weather is designed on the same layer principle as cold weather clothing. It is, however, made of only two layers. The outside layer is made of strong, tightly woven fabric. It is water repellent and impervious to rain and snow. The inner layer is much lighter in weight than the outer layer. The liner is fastened by toggles to the tent and provides an air space the same as in clothing. It is designed to provide insulation against the cold. It also prevents frost from forming on the inside of the tent. Heat is provided by stoves (normally the Yukon stove).

b. Tent, Arctic, 10-Man (fig. 9). The six-sided, pyramidal tent, supported by a telescopic pole, normally accommodates ten men and their individual clothing and equipment. In an emergency it may accommodate up to fifteen men by leaving their individual packs and equipment ouside the tent overnight. It may also function as a command post, aid station, or as a small storage tent. The tent has two doors; this permits tents to be joined together, with access from one to the other, when additional space is required. A snow cloth is attached to the bottom of the side walls for sealing around the bottom of the tent. This is accomplished by piling and packing snow on the snow cloth. If the tent is used in terrain where there is no snow, sod or other materials may be used to seal the bottom of the tent. Flexible plastic screen doors are provided and may be attached to the front and rear of the tent for protection against insects. The tent is ventilated by four built-in ventilators on opposite sides and near the peak of the tent. Four lines are provided for drying clothing and equipment. Total weight, to include the pins and tent pole, is 76 pounds. The tent is heated by an M1950 Yukon stove.

c. Tent, Hexagonal, Lightweight (fig. 10). This tent is also six-sided, pyramidal, and supported by a telescopic tent pole. It is

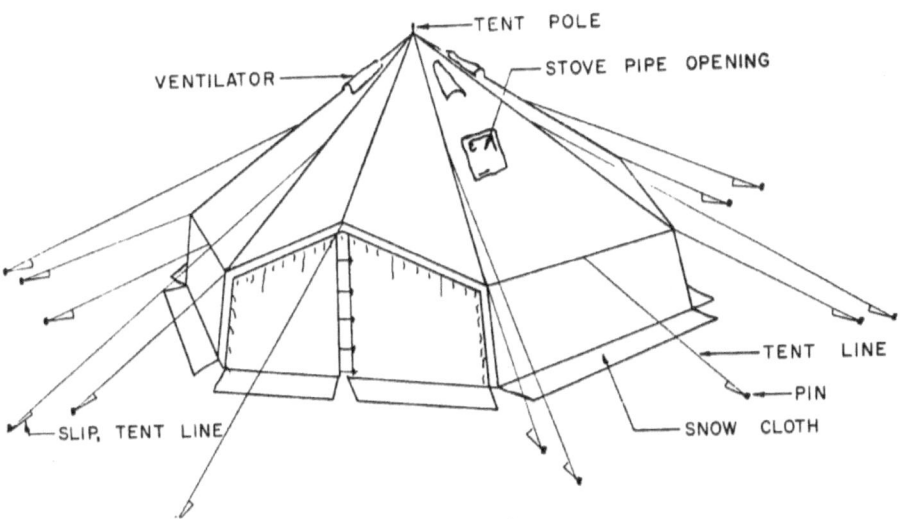

Figure 9. Tent, arctic, 10-man.

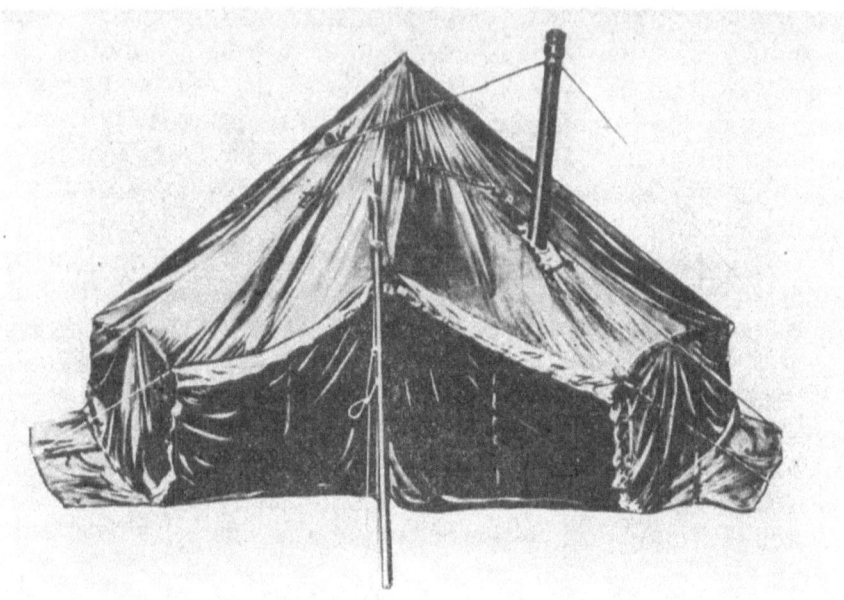

Figure 10. Tent, hexagonal, lightweight.

designed to accommodate four to five men and their individual clothing and equipment. Under emergency conditions one tent may provide shelter for a rifle squad or other similar unit when rucksacks are placed outside the tent. The tent has one door; ventilation is provided by two built-in ventilators located on opposite sides and near the peak of the tent. Three lines are provided for drying clothing and equipment. Total weight of the tent, including the pins and center poles, is 48 pounds. The tent is heated by an M1950 Yukon stove.

d. Tent, Frame-Type, Sectional (Jamesway). This 16x16 frame-type tent (fig. 11) is a lightweight unit that offers protection for one squad. It has wooden floor units, a frame, a rounded roof, and comfortable head clearances along the center line of the shelter. The roof and ends of the tent are fabricated from insulated, coated, fabric blankets. The structure is fastened to the ground with tent pins or on snow with improvised devices. It weighs approximately 2,250 pounds and is heated by one tent stove M1941. The heavier weight of this tent restricts its normal use to permanent or semipermanent base camps. It could be used by forward elements under stabilized conditions.

28. Pitching and Striking Arctic Tent

a. With proper training, small troop units will be able to pitch the tent in 15 to 20 minutes. Additional time will be required to

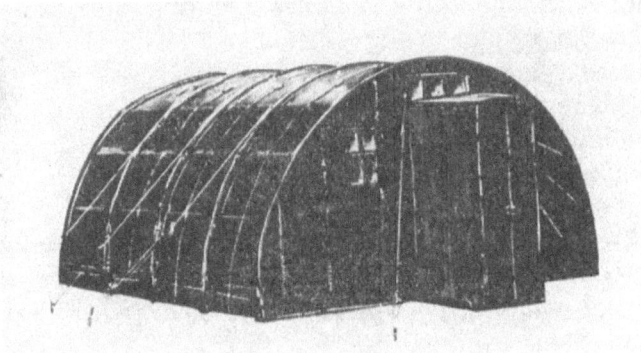

Figure 11. Tent, frame type, sectional.

complete the camouflage of the tent. Pitching and striking of the tent are performed in a routine drill manner in accordance with instructions contained in FM 20-15.

b. The following must be considered when pitching or striking the tent in snow or on frozen ground:

   (1) Snow should be cleared to one foot from ground level to provide camouflage, protection from the wind, and to decrease the danger from hostile small arms fire. The remaining foot of snow should be tramped down to act as an insulator against the cold ground. An adequate tent site may be packed on snow by stamping with skis and snowshoes until a firm base is provided for pitching. In this case, the tent pole is placed on a log or other suitable support to keep the pole from sinking into the snow. Support is also needed for the stove under similar conditions.

   (2) In open terrain, with a strong wind, it may become necessary to build a snow wall on the windward side of the tent to protect it from the wind. The snow wall also makes it easier to heat and less likely to blow down. The tent is pitched with the entrance downwind. Variable winds may require construction of a wind break at the entrance. High winds in certain cold areas necessitate anchoring the tent securely. When the tent is set up, the snow cloth should be flat on the ground outside the tent. Stones, logs, or other heavy objects should be placed on the snow cloth in addition to the snow to assist in anchoring the tent. If this is not done, the tent will be drafty and very difficult to keep warm.

   (3) Tents may be pitched rapidly and anchored securely by attaching the tent lines to trees, branches, logs or stumps whenever possible. If these natural anchors are

not available, suitable holes are dug into the snow for the purpose of using "deadmen." This is accomplished by digging a hole into the snow large enough to insert a pole or log approximately 2 to 4 feet long with the tent line attached. The hole is then filled with snow, well packed, and in a short period of time the packed snow freezes and the tent will be securely anchored (a, fig. 12). Driving metal pins into frozen or rocky ground should be avoided when excessive force is required. On rocky ground, tent lines may be tied around heavy rocks and then weighted down with other stones (b, fig. 12).

(4) Tents are also occasionally pitched on ice. When the thickness of the ice is not excessive, a small hole is chopped through the ice. A short stick or pole with a piece of rope or wire tied in the middle of it is pushed through and then turned across the hole underneath the ice (c, fig. 12). If the ice is very thick a hole 1 to 2 feet deep is cut in it, the "deadman" inserted and the hole filled with slush or water (d, fig. 12). When the slush or water is frozen, an excellent anchor point is provided. When the "deadman" is placed underneath or into the ice, a piece of rope or wire should be fastened to the "deadman" and the tent line then fastened to the rope or wire after the "deadman" is secure. This may prevent the tent line from being accidently cut or damaged when being removed from the ice.

(5) When striking the tent in winter it will normally be covered with snow and ice which must be removed or the tent may double in weight. Snow and ice can easily be removed by shaking the tent or by beating it with a

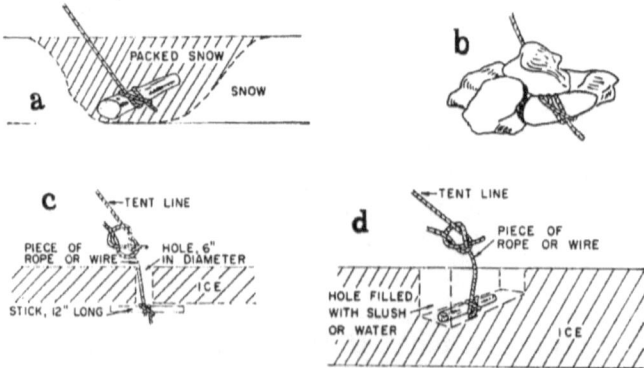

a. *"Deadman" buried in packed snow.*
b. *Tent line buried under stones.*
c. *Lowering anchor through ice.*
d. *"Deadman" placed in thick ice.*

Figure 12. Improvised methods of anchoring tents.

mitten or a stick. If the snow cloth is frozen to the ground, the snow and ice around it must be carefully removed by chopping or shoveling in order to avoid damage to the material. One method of accomplishing this is to ease the shovel between the cloth and the ground and gently pry the cloth away from the ice.

29. Ventilation

a. Tents are pitched to protect occupants from the elements and to provide necessary warmth and comfort. When the bottom of the arctic tent is properly sealed and the doors are zipped shut, the tent becomes almost airtight and very little fresh air can enter. Moisture will form on the inside of the tent and accumulate on clothing and equipment, thereby causing dampness and hoarfrost. In addition, carbon monoxide, carbon dioxide, and fumes from the stoves may soon accumulate to a dangerous degree. To offset these factors, the built-in ventilators near the peak of the tent must be kept open.

b. To improve ventilation, a draft channel may be constructed by forming a pipe with green logs (fig. 13). The channel is buried in the floor and has an opening under the stove. The draft of the stove draws fresh air from outside the tent into the channel.

30. Heating Tents With Yukon Stove

a. Stove, Yukon, M1950. The Yukon stove (1, fig. 14) is used with both the arctic 10-man and the hexagonal 5-man tent. In addition to providing heat, the top surface of the stove and, to a small degree, the area beneath the stove, may be used to cook rations or heat water. The Yukon stove utilizes standard leaded motor fuel as its normal fuel, but may also be operated with white gasoline, kerosene, light fuel oil, naptha, or jet fuel, without modification (2, fig. 14). When solid fuels are used, wood, coal, etc., the stove

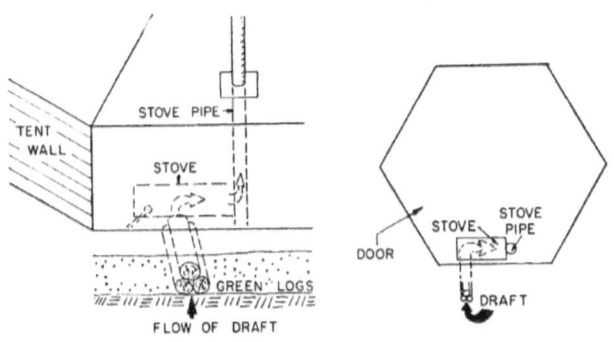

Figure 13. Draft channel for stove.

must be modified by removing the oil burner from the top of stove, closing the opening where the burner was installed, and turning over the wire grate so that there is space below grate for ashes. It is also necessary to *remove* the draft diverter and to install a spark arrester either on top of stovepipe or between the first and second pipe sections of stovepipe (counting from the bottom). The standard 4-inch stovepipe can be applied to the top of stovepipe outside the tent only. For figuring consumption, five gallons of gasoline will last approximately eight to twelve hours in low temperatures.

b. Operating Procedures. The compact, lightly constructed, 33-pound Yukon stove permits all accessory parts to be packed within the stove body for convenient portability in a man-hauled sled or on a packboard. A draft diverter is issued as a component part of the stove. It shields the top of the stovepipe from the wind and prevents backdraft from forcing smoke or gasses into the stove and tent. Three 15-foot guy lines serve to anchor the draft diverter against strong winds. A simple method of erecting a tripod for a fuel can is to obtain three poles about 6 feet in length; the poles are tied about two-thirds of the way up using wire from ration cases, string, rope, or emergency thong, and then spread out to form a tepee. The fuel can should be at least 3 to 4 feet higher than the stove. The lowest part of the inverted gasoline can should be a minimum of one foot above the level of the needle valve of the Yukon stove. It should not be higher than four feet if the valve is to operate smoothly. If the fuel can is wobbly or if there is some wind the can must be tied to the tripod for additional protection. Make certain that the can is tilted so that air is trapped in the uppermost corner. The stove itself is assembled, operated, and maintained in accordance with TM 10-735.

c. Precautions. The following precautions must be observed when the Yukon stove is used:

  (1) *Burning liquid fuels:*

   (a) All stovepipe connections must be tight and necessary tent shields adjusted properly.

   (b) Stove must be level to insure that the burner assembly will spread an even flame within the stove.

   (c) The fuel hose must be protected so it cannot be pulled loose accidently. If necessary, a small trench may be dug and the hose imbedded where it crosses the tent floor.

   (d) The fuel line must not be allowed to touch the hot stove.

   (e) When adjusting the fuel flow, the drip valve lever must be turned carefully to prevent damage to the threads.

(f) Rate of fuel flow must be checked at regular intervals. The rate of flow will change as fuel supply level drops and will require some adjustment. The stove should never be left unattended. Maintaining a hotter fire than necessary may cause the stove body to become overheated and warp.

(g) If the flame is accidently extinguished, or if the fuel can is being changed, the drip valve must be closed. When the stove has cooled, any excess fuel inside the stove must be wiped up and two or three minutes allowed for gas fumes to escape before relighting the burner. The burner must be cool before relighting stove. If stove is lit before burner is cool, the fuel will vaporize prior to ignition, causing an explosion.

(h) All fuel supplies must be kept outside the tent. Spare cans of gasoline or other fuel should never be stored inside the tent. Fuels used in combat areas in the North are normally low temperature fuels which will flow freely.

(2) *Burning solid fuels:*
   (a) A small amount of fuel should be fed at a time until the bed of coals is burning brightly.
   (b) Stove should not be allowed to overheat.
   (c) Oil or gasoline should not be poured on the fire.
   (d) Ashes should not be allowed to accumulate below the grate.
   (e) Clinkers should be removed to prevent grate from becoming blocked.
   (f) Spark arrester must be installed between the first and second pipe sections when solid fuel is used, to prevent sparks from falling on the tent and burning holes. The spark arrester is not a component part of the stove but may be requisitioned separately. The diverter and arrester will never be used at the same time.

31. Heating of Semipermanent Tents With Tent Stove, M1941

Stoves of this type are normally used to provide heat for the semipermanent, frame-type, sectional tent. The stove may be operated with wood or coal or with various types of oil and gasoline. This stove has the same general characteristics and safety features outlined for the Yukon stove in paragraph 28.

32. Fuel Economy

The minimum daily fuel consumption per stove approximates five gallons of gasoline per eight to twelve hours of operation.

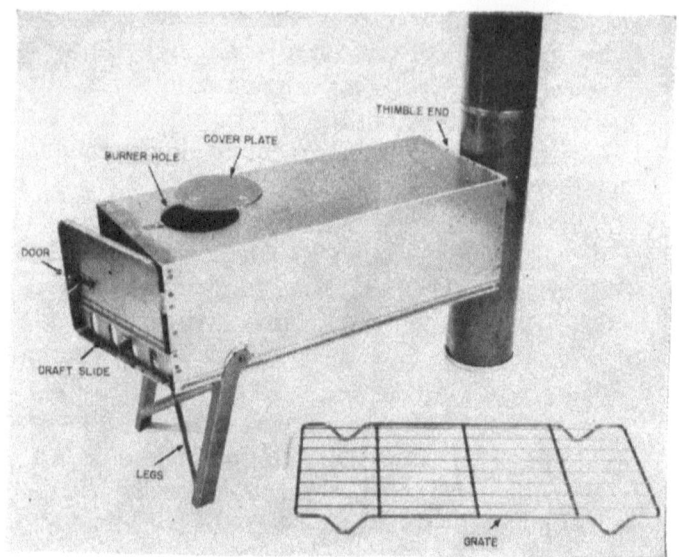

1—*Body assembly.*

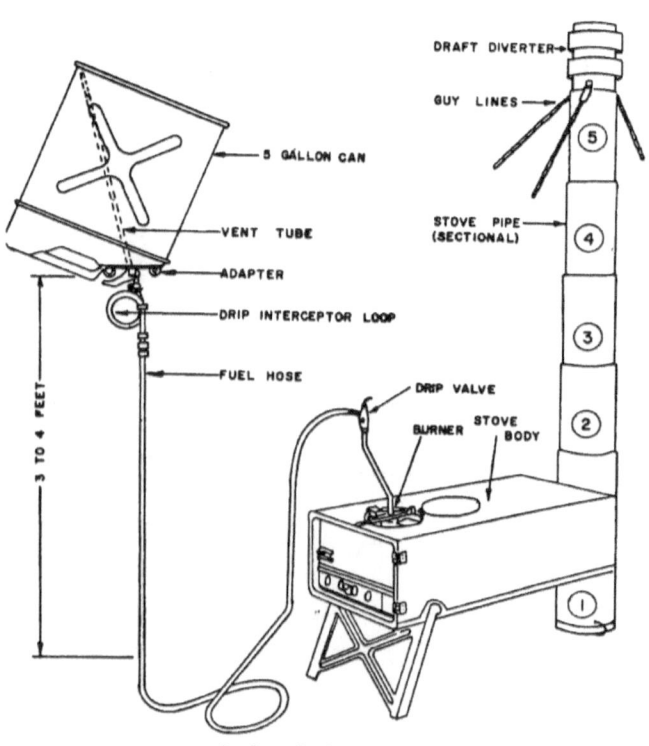

2—*Installation set-up.*

Figure 14. Stove, Yukon, M1950.

Prior planning must be accomplished to reduce the number of stoves required, especially for operations that are some distance from a road net. Wood should be used as fuel whenever possible. Cooking and heating are combined and, when extra heat is required to dry clothes, all individuals should dry clothes at the same time, when possible.

33. Lighting Tents

Candles will provide light in forward areas. In rear areas, gasoline lanterns or lighting equipment sets may be used.

34. Tools

a. In addition to the tent, tent stove, and its accessories, tools are needed by small units for several purposes, such as erecting and striking the shelter, building ski and weapon racks, building field latrines, chopping ice holes, chopping fire wood, etc. Tools are also needed for trailbreaking, preparation of positions, and similar tasks. As intrenching tools are lightly constructed, they are of little value for work in heavy timber or frozen ground. The following tools are needed by small units to accomplish routine tasks in cold regions, regardless of the season of the year:

(1) One axe, chopping.
(2) One saw, (buck or swede).
(3) Two machetes with sheaths.

b. Well trained individuals will maintain their tools at all times. Tools must be kept sharp, clean, oiled, and otherwise in good shape. Tools should never be left in the snow or thrown aside where they may become easily lost.

c. Particular care must be exercised while wearing gloves, as ice or frost may form on the gloves and cause the tools to slip from the user's hand.

Section III. IMPROVISED SHELTERS

35. Requirement for Improvised Shelters

a. There are many occasions when tents or other regular shelters are not available. In summer, if the weather is mild, individuals may need protection only from insects. In winter, however, individuals cannot stay in the open for long periods unless they are moving. The requirement for improvised shelters may arise for several reasons, e.g., vehicles carrying tents may be unable to reach the troops due to difficult terrain or enemy action. In case of emergency, each individual must know how to protect himself from the effects of the weather.

b. If suitable natural shelters such as caves or rock shelves are available, they should be used. If natural shelters are not available, a temporary improvised shelter must be established.

c. The type of improvised shelter to be built depends on the equipment and materials available. By the proper use of materials available, some sort of shelter can be built during any season of the year. In open terrain a shelter can be built using ponchos, canvas, snow blocks, or other materials. Snow caves, snow trenches, or snow holes may be constructed in the winter if the snow is both deep and well-compacted. In the woods, a lean-to is normally preferable to other types of shelter. In the North, nature provides the individual with the means to prepare a shelter. His comfort, however, greatly depends on his initiative and skill at improvising.

d. Construction techniques of various types of improvised shelters are discussed in paragraphs 36 through 42.

36. Poncho Shelters

A poncho is a part of an individual's cold-wet weather uniform. It is a multipurpose piece of equipment and may be used as a rain garment, a waterproof bed cover, or a shelter. The simplest type of shelter can be made by merely pulling the poncho over the sleeping bag. For additional comfort, various types of shelters and lean-tos may be made by attaching ponchos to trees, tree branches or poles.

a. One-Man Lean-To. A one-man lean-to (1, fig. 15) may be made from one poncho. The poncho is spread, hood side up, on the ground, and the hood opening is tightly closed by adjusting and tying the hood draw-strings. The poncho is raised at the middle of its short dimension to form a ridge, and then staked out at the corners and sides. Side stakes should not be driven through the grommets at the corners or sides, as this may tear the poncho. A short piece of rope is tied to the grommets and, in turn, to the stakes.

b. Two-Man Lean-To. To construct a two-man lean-to (2, fig. 15), ponchos are spread on the ground, hood side up, with the long sides together so that the snap fastener studs of one poncho may be fastened to the snap fastener sockets of the other poncho. Hood openings must be tightly closed by adjusting and tying the hood draw-strings. Ponchos are raised where they are joined to form a ridge; ropes are then attached to grommets at the ends of the ridge and run over forked sticks. The shelter tent is then staked out at the corners and sides, as described in *a* above. A third poncho may be snapped into the other ponchos to form a ground cloth.

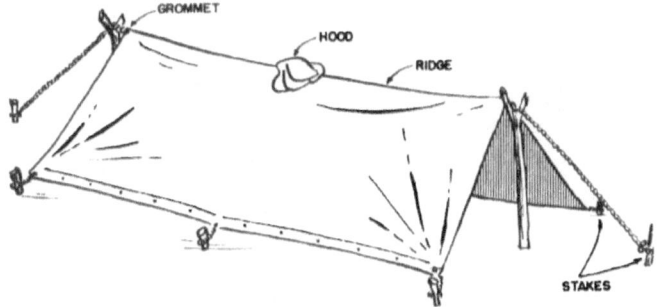

1. One-man lean-to.

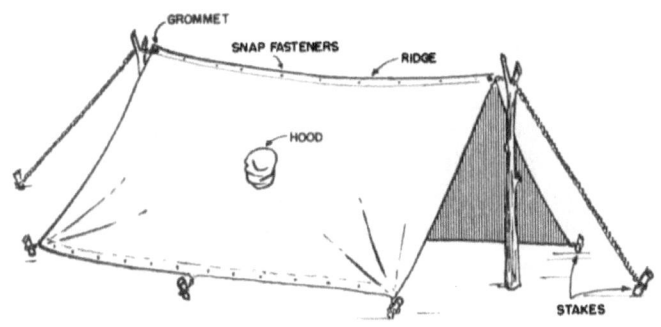

2. Two-man lean-to.

Figure 15. Poncho shelters.

 c. Insect Bar, Nylon Netting, Field Type. A standardized field-type insect bar is a canopy made from netting of small nylon mesh. It is used to afford protection against insects under summer conditions. When a tent is not available, or is not being used, the insect bar is suspended by tying the tie tapes at the top corners to trees or bushes. Using poncho shelters, the insect bar is fastened inside the two-man shelter in a manner similar to fastening it to the shelter half tent (fig. 16).

37. Lean-To.

 a. Materials. The lean-to shelter, used in forested areas, is constructed of trees and tree limbs. String or wire helps in the building, but is not necessary. A poncho, a piece of canvas, tarpaulin, or a parachute, in addition to the boughs, may be used for covering.

 b. Size. The lean-to is made to accommodate a variable number of individuals. It may be built for one man only, teams, gun crews, patrols, or similar small groups. From a practical point of view, a rifle squad is the largest element to be sheltered in one double lean-to.

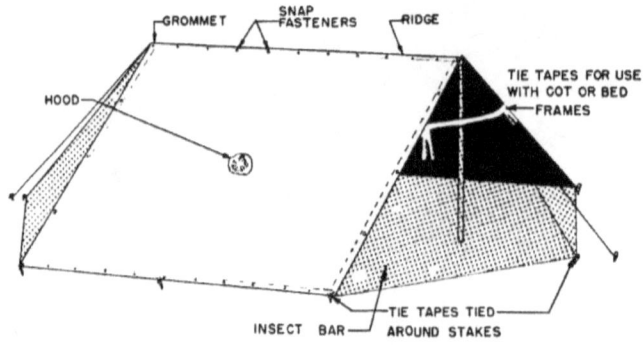

Figure 16. Using field-type insect bar with two-man poncho shelter.

c. *Types.* Depending on the number of individuals to be sheltered, two types of lean-tos, single and double, are used.

d. *Construction.*

(1) *Single lean-to* (fig. 17). To save time and energy, two trees of appropriate distance apart, and sturdy enough to support the crosspiece approximately five feet off the ground, are selected when operating in forested areas. It may be necessary to cut two forked poles of desired height, or construct two "A" frames to hold the crosspieces, or use a combination of these supports when bivouacking in sparse wooded or semi-open areas. A large log is then placed to the rear of the lean-to for added height. Other methods that may be used are packing the snow down or using snow blocks instead of a heavy log. Stringers approximately ten feet long and two to three inches in diameter are then placed, approximately eighteen inches apart, from the crosspiece over the top of the log in the rear of the shelter. One or both sides of the lean-to, and the roof, are then thatched. In some cases it may be necessary to line the inside of this type shelter with ponchos or pieces of cardboard to keep melting snow, warmed by the fire, from dropping through.

(2) *Double lean-to* (fig. 18). Two single lean-tos are built facing each other and approximately five to six feet apart. The space between single lean-tos must be sufficient to permit the occupants to move freely around the log fire placed along the center line of this space and to allow the smoke to get out through the opening instead of gathering under the roofing. If desired, one end of the middle space may be covered by a wall made of boughs or other materials for additional protection from the draft and wind.

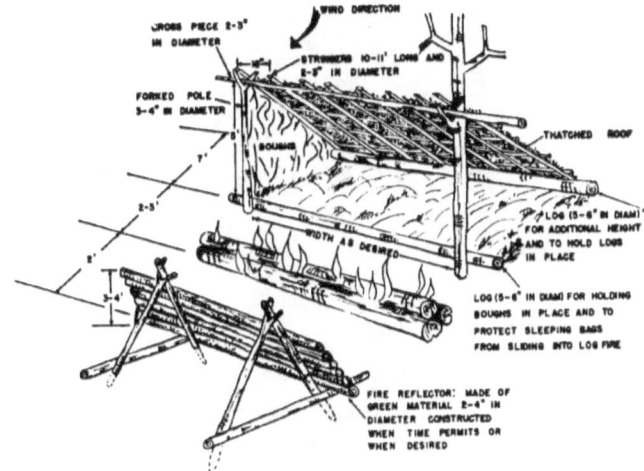

Figure 17. Single lean-to.

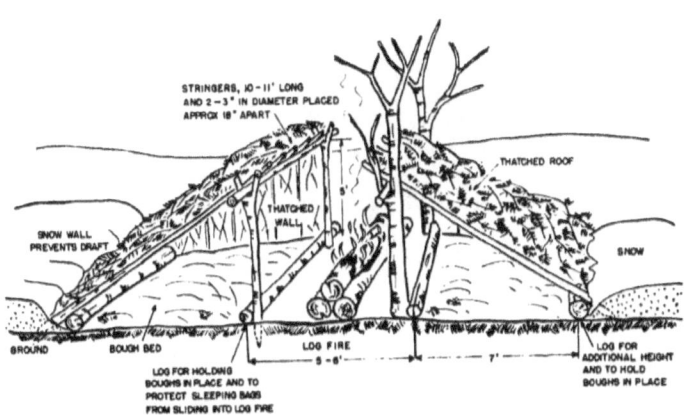

Figure 18. Double lean-to.

e. Heating. In heating a lean-to, any kind of open fire may be used. The best type for large size lean-to, however, is the log fire, so the heat will be evenly distributed over the entire length of the lean-to.

38. Tree-Pit Shelter

In wooded areas, the deep snow and tree-pit shelter (fig. 19) furnishes temporary protection. To construct a tree-pit shelter a large tree is selected with thick lower branches and surrounded with deep snow. The snow is shaken from the lower branches and the natural pit is enlarged around the trunk of the tree. The walls and floor are then lined with branches and the roof thickened. Canvas or other material on hand may be used for the roof.

Figure 19. Tree-pit shelter.

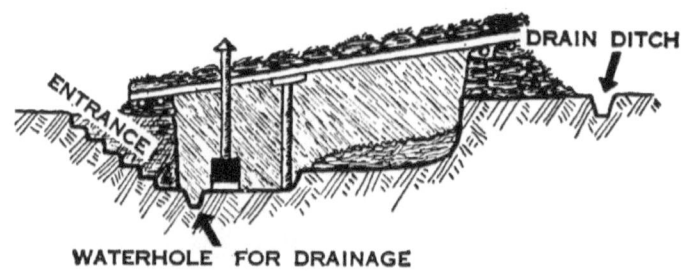

Figure 20. Sod house.

39. Sod House

A sod house (fig. 20) should be located on level ground with good drainage. To provide a watertight roof, ponchos, birch bark, cardboard, or any other available material should be used in addition to the sod.

40. Snow Wall

In open terrain with snow and ice, a snow wall (fig. 21) may be constructed for protection from strong winds. Blocks of compact snow or ice are used to form a windbreak.

41. Snow Hole

A snow hole (fig. 22) provides shelter quickly. It is constructed by burrowing into a snow drift or by digging a trench in the snow and making a roof of ponchos and ice or snow blocks supported by ski poles or snowshoes. A sled provides excellent insulation for the sleeping bag. Boughs, if available, can be used for covering the roof and for the bed.

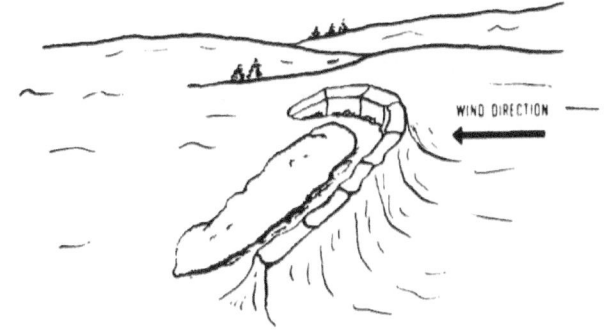

Figure 21. Sleeping behind snow wall.

1. *Snow hole partially constructed.*

2. *Snow hole completed.*

Figure 22. Snow hole.

42. Snow Cave

a. Location. A snow cave (figs. 23 and 24) can be used as an improvised shelter in the open areas where deep and compacted snow is available. Normally, a suitable site is located on the lee side of a steep ridge or river bank where drifted snow overhangs in unusual depths.

b. Basic Construction Principles. Basic principles for construction of all snow caves are as follows:

(1) The tunnel entrance must give access to the lowest level of the chamber, which is the bottom of the pit where cooking is done and equipment is stored.

(2) The snow cave must be high enough to provide comfortable sitting space.

(3) The sleeping areas must be on a higher level than the highest point of the tunnel entrance so that the rising warm air will permit the men to sleep more comfortably.

(4) The roof must be arched both for strength and so that drops of water forming on the inside will not fall on the floor, but will follow along the curved sides, glazing over the walls when frozen.

(5) The roof must be at least one foot thick.

c. Size. The size of the snow cave depends upon the number of men expected to occupy it. A large cave is usually warmer and

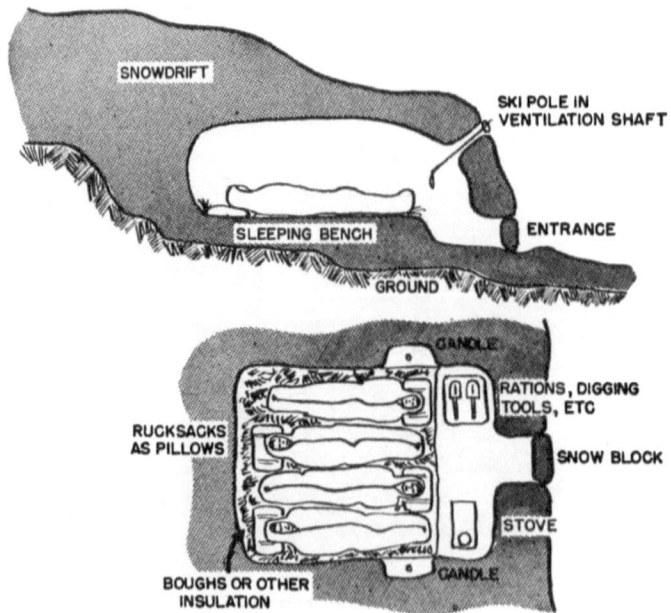

Figure 23. Snow cave for four men.

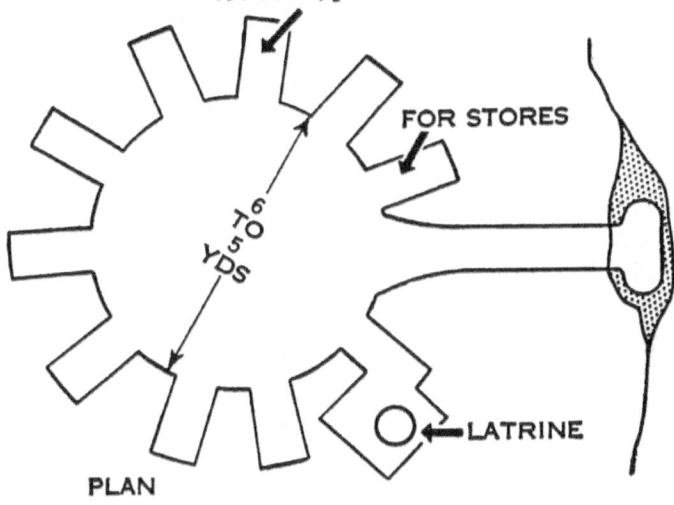

Figure 24. Snow cave for sixteen men.

more practical to construct and maintain than several small caves. In good snow conditions a 16- to 20-man cave is the most practical.

d. Shape. The shape of the snow cave can be varied to suit conditions. When the main cave is built, short side tunnels are dug to make one-or-two-man sleeping rooms, storage space, latrine and kitchen space.

e. Construction. The following steps should be observed in construction:

(1) A deep snow drift is located. Newly fallen, powdery, or too heavily packed snow should be avoided.

(2) The depth of the snow can be tested with a ski pole.

(3) The entrance is chosen carefully so the wind will not blow into the cave or the entrance become blocked by drifting snow.

(4) A small tunnel is burrowed directly into the side of the drift for two feet. A chamber is excavated from this tunnel.

(5) Excavation is done to the right and left so that the length of the chamber is at right angles to the tunnel entrance.

(6) Due to the fact that individuals digging will become wet, they should wear the minimum amount of clothing possible to insure that they have a dry change of clothing upon completion of the task.

f. Heating and Safety Measures. The Yukon Stove provides the best means for heating the snow cave. Weather permitting, the stove should be turned off when individuals are sleeping, thus reducing the danger of fire and asphyxiation. If the weather is severe and it becomes necessary to keep a fire going while the individuals are sleeping, an alert fire guard must be posted in each shelter and safety precautions as outlined in paragraph 30c strictly adhered to.

g. Insulation. To insure that the cave is warm, the entrance should be blocked with a rucksack, piece of canvas, or snow block when not in use. All available material, such as ponchos, cardboard, brush, boughs, etc., should be used for ground insulation.

h. Other Precautions. Walking on the roof may cause it to collapse. At least two ventilators, one in the door and one in the roof, are used. A ski pole can be stuck through the roof ventilator to clear it from the inside. Extra care must be exercised to keep air in the cave fresh when cooking.

43. Camp Fires

a. Matches and Fire Starters. A supply of matches in a waterproof container, heat tablets, or fire starters must be carried by all individuals operating in cold weather. They are a necessity, especially where snow and ice add to the problems of securing tinder for starting a fire. In emergencies, matches should be used sparingly and lighted candles used to start fires whenever possible.

b. Selecting Site. Individuals building a fire in the field should carefully select a site where the fire is protected from the wind. Standing timber or brush makes a good windbreak in wooded areas, but in open country some form of protection must be provided. A row of snow blocks, the shelter of a ridge, or a scooped-out side of a snow drift will serve as a windbreak on barren terrain.

c. Starting and Maintaining Fire. Before using matches, a supply of tinder must be on hand. The use of fire starters, heat tablets, or pieces of paper or rag soaked with gasoline will make fire starting easier. Unfortunately, these items are not always available and individuals must use other materials to start a fire. Many types of fuel are available for fires. The driest wood is found in dead, standing trees. Fallen timber may often be wet and less suitable. In living trees, branches above snow level are the driest. Green and frozen trees are generally not suitable because they will not burn freely. Splitting green willows or birches into small pieces provides a fairly good method of starting and maintaining a fire, if no dead wood is available. Also, dry grass, birch bark, and splits of spruce bark with pitch tar are excellent fire starters. It is good practice to secure a sufficient amount of fire wood to last throughout the night, before retiring.

d. Types of Fire. Any kind of open fire may be used with most of the improvised shelters. In deep snow, a fire base (fig. 25) of green wood should be built first to protect the camp fire from sinking into the snow. For a single lean-to or snow wall, a fire reflector may be built of green logs or poles to reflect the heat into the shelter and to serve as a windbreak, allowing the fire to burn more steadily. The most suitable types for single and double lean-tos are the log fires (fig. 26).

   (1) Two, preferably three, logs are used for this type of camp fire. Dry logs, 8 to 15 inches in diameter and approximately the same length as the lean-to, are selected and brought to the fire site. First, two logs are placed side by side on small green blocks to support them above the snow or ground for a better draft. Then the third log is placed in the middle and on the top of the other two logs. For better burning, the surfaces of logs which face each other are chipped. Before lighting the fire, small wedges are placed between the chipped surfaces of the logs for better draft. Fire is then started at several places to help it spread the entire length of the logs.

   (2) When only two logs are used, four vertical stakes must be driven into the snow to keep one log on top of the other.

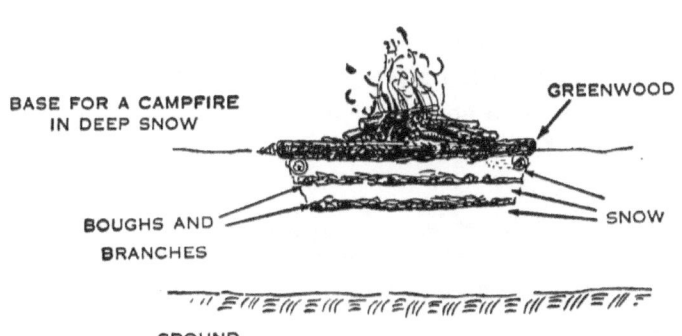

Figure 25. Camp fire and fire base on snow.

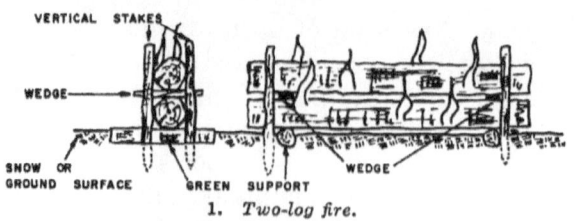

1. *Two-log fire.*

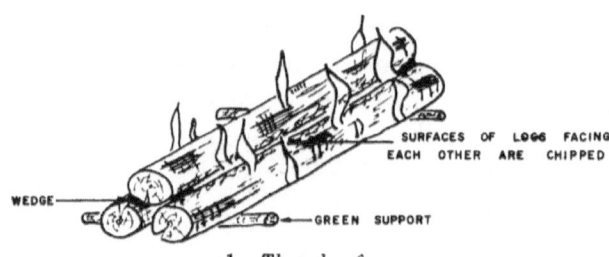

1. *Three-log fire.*

Figure 26. Log fires.

A disadvantage of this type of log fire is the fact that the vertical stakes tend to give away when the snow starts melting around the fire. A log fire will normally burn all night, requiring only a minimum of care.

Section IV. FOOD AND WATER

44. Principles

a. Importance of Balanced Meals. Army rations are well balanced. The ration for 1 day provides all the essential foods the body requires. However, all the ration must be eaten if all the caloric value is to be obtained. Some items may, at times, not appeal to the individual sense of taste, but they must be eaten. The tendency to be lazy about preparing and eating satisfactory morning and evening meals before and after a hard day on the trail must be avoided, since it is exceedingly detrimental to continued good health. After having been without normal supplies for a period of time, it is essential that men be provided with a balanced meal containing the three basic food requirements (fats, protein, and carbohydrates).

b. Importance of Liquids. In cold regions, as elsewhere, the body will not operate efficiently without adequate water. Dehydration, with its accompanying loss of efficiency, can be prevented by taking fluids with all meals, and between meals if possible. Hot drinks are preferable to cold drinks in low temperatures since they warm the body in addition to providing needed liquids. Alcoholic beverages should not be consumed during cold weather operations.

45. Rations

Many types of rations will be used for operations in cold weather. There may be rations of varying components prepared in unit kitchens, rations issued to, and prepared by, the individual, or rations which are issued boxed and prepared for use by the small detachment. The type of ration to be used will be determined by the location, supply situation, mission, and the duration of the trip or stay. When at all possible, rations are prepared in the unit kitchens. In situations where this is not practicable, group rations are generally utilized and prepared by one member of the small unit. In certain conditions an individual ration may be issued to each man. Whenever possible, shelters should be provided for the preparation and serving of food. During the winter, without shelter, food becomes cold before it can be eaten. Emergency rations are precooked and can be eaten without heating, but are more palatable when heated.

a. "A" Rations. The standard "A" ration of fresh foods will be issued whenever possible. The caloric content of the ration has been increased to compensate for the added rigors of cold weather. The "A" ration, with its variety of components, allows for a large variety in menus. Troops will be supplied with "A" rations while in garrison, under stablized conditions in the rear areas, and in forward areas whenever possible.

b. Packaged Group Rations. When the situation does not allow the issue of fresh perishable rations, a packed non-perishable ration will be provided. This ration is more compact than the fresh ration, but does not provide the same variety of foods.

c. Frozen Rations. Frozen rations consist of pre-cooked frozen foods prepared at a central kitchen, to be thawed and heated by the forward troops after delivery. These frozen rations give greater variety, are more palatable, and easily prepared. At the present time they are limited to winter use because of the difficulties of storage.

46. Individual Rations

a. The small detachment ration, such as the five-in-one type, consists of packages of prepared food. If no heating facilities are available, it may be eaten cold. To avoid monotony, several menus should be issued when it is not practical to use fresh rations.

b. The individual combat ration consists of packages of prepared or precooked food. The principal purpose of this ration is to provide each man with a personal ration allowance while under field conditions, when cooking facilities cannot be made available. The ration should be heated whenever possible, using any means available.

47. Food Packets

These are individual sources of nourishment for use in specific operational situations. They consist of precooked or prepared foods, especially selected for the maximum nutritional value, palatability, and utility, compatible with minimum weight and bulk. They are considered as an emergency or special purpose item. The two types most used in cold areas are:

a. Food packet, assault, individual, to be used in the initial assault phase of combat.

b. Food packet, survival, arctic, to be carried aboard aircraft for use in event of emergency. This type may also be used by personnel participating in active operations (in an emergency). Each packet contains food for one man for one day.

48. Individual or Small Unit Messing

Frequently, while on patrol or during combat conditions, individuals will find it necessary to prepare their own meals or to combine rations with other individuals within the unit.

a. Equipment. Any fuel-burning device will give off carbon monoxide gas, which is poisonous. Adequate ventilation must be provided when using the equipment under shelter.

 (1) The one-burner M1950 gasoline cooking stove is a cooking and heating unit for a group of from 2 to 5 men operating in an isolated or forward area where the use of heavier equipment is not practical. The mountain cookset is combined with the stove to make the one-burner cooking outfit. Detailed instructions for operation and maintenance are contained in TM 10–708.

 (2) The small detachment cooking outfit consists of a gasoline or kerosene-burning stove and the necessary attachments and utensils required to prepare rations for 15 to 40 men. This outfit is designed primarily for outdoor use by isolated detachments. Detailed instructions for operations and maintenance are contained in TM 10–703.

b. Preparation.

 (1) First priority is the procurement of water (par. 55). If

snow or ice must be melted to obtain water, all available stoves are utilized for this purpose. After water is obtained, the stoves are used for food preparation. For convenience in preparation of meals and for conservation of fuel and labor, cooking should be done for as large a group as the situation permits.

(2) Meals must be prepared quickly and efficiently. Areas sheltered from the wind should be chosen for stoves or fires. A few blocks of snow or ice or a hole dug in the snow will serve as a windbreak and provide for more efficient use of fires. Heating tablets are not adequate in extremely cold weather accompanied by high winds. Individuals may have to prepare and eat one item at a time, but a hot meal will be worth the effort.

(3) Instructions for preparing the components of the rations will be found on, or inside, the package. The possibility of combining the various ration components, i.e., mixing meat and vegetables to make stew, should also be considered.

(4) Canned foods are cooked and require little heat to make them edible. Overcooking will waste fuel. The juices in canned vegetables are tasty, and contain vitamins and minerals. Drinking them will conserve the water supply. Cans must be punctured or opened before heating by open fires or stoves. Failure to do this may result in an explosion. No puncturing is needed if the can is submerged in water during the heating process.

(5) Food, including frozen meat, should be thawed before cooking. Partly frozen meats may cook on the outside while the center remains raw. Fresh meats must be cooked thoroughly to kill any germs or parasites that may be present.

(6) Whenever possible, dried fruit should be soaked overnight in cold water, then simmered slowly in the same water until tender, and sweetened to taste.

c. Storage.

(1) In winter the simplest way to preserve perishable foods is to allow them to freeze. Rations should be stacked outside the shelter and their location carefully marked. Only as much food as can be thawed and consumed before spoiling should be brought into the shelter.

(2) Frozen food should not be placed near heat where it may be thawed and later refrozen. Once thawed, certain foods may spoil. Meat thawed and refrozen two or three times

is tasteless and watery, and resultant bacterial growth may be sufficient to cause food poisoning.

d. Eating. Meals should be prepared at regular times and as much time as possible allowed for cooking and eating. Men should be allowed to relax after each meal. There will be times when it may not be possible to prepare a lunch. Under such circumstances the components for lunch must be distributed to the men at breakfast. Any frozen food may be thawed before breaking camp in the morning. The men can carry it under their shirts to keep it from refreezing. By this method, considerable time and fuel is saved and, if it is not practicable to have a fire, the food can still be eaten. If time permits, the midday stop should be long enough to prepare hot food or at least a hot drink. All possible preparation of the following day's food should be done during the night stop in order to shorten the time required to break camp in the morning.

e. Suggestions.
  (1) Organize and control cooking.
  (2) Insure that all food is eaten; save any usable leftovers for snacks between meals.
  (3) The squad leader supervises the meals and makes sure that each man is receiving his portion.
  (4) Check continuously to see that each man's mess equipment is kept clean.
  (5) Food is prepared for as large a group as possible.
  (6) Fuel is conserved by prethawing food. This may be done by utilizing heat in the engine compartment of a vehicle or by placing cans of food under and around the tent heating stove.
  (7) Canned rations, either frozen or thawed, can best be heated by immersion in a pot of hot water on the stove. This water can then be used for washing soiled utensils.
  (8) Adequate training of all men in the preparation and cooking of cold weather rations is imperative.
  (9) One-pot meals, such as stews, save preparation time and fuel and can be kept warm more easily than several different food items.

49. Small Unit Messing

a. One Man Responsible. One man should be responsible for the preparation of each meal and this job should be rotated throughout the squad. The squad leader is responsible for supplying any additional assistance needed by the cook.

b. Ingenuity in Cooking. Ingenuity on the part of the man assigned to cook for the small unit will aid immeasurably in the success of field messing in cold weather. Potatoes, onions, or bacon, when available, will increase the palatability of the food and can satisfactorily be added to many foods. The habit of making the morning coffee the night before, or using two stoves to melt snow or ice for the evening's water supply, and of thawing out those rations that are going to be used the next morning, will save time and greatly simplify food preparation at meal time.

c. Eating Arrangement. When the weather is moderate, the mess line feeding system may be used. During cold weather in a bivouac area the food can be prepared hot and then carried in insulated containers to each tent for consumption in a heated shelter. Food may also be transported in this manner to front line troops by using track vehicles or other methods of transport.

50. Natural Food Resources

a. In some cold regions, edible plants and animals are abundant at certain seasons of the year. In other areas, very little game or edible plant life can be found during any season of the year. A person without food in these areas must know how to "live off the land" and subsist on what is available. Fish are present in fresh water lakes and rivers during all seasons of the year, and some salt water near shore will normally yield fish. Fish will form the most readily available and largest portion of available nourishing foods.

b. Small animals and birds are also present in most areas at all times of the year. Large animals, because of migratory habits or other characteristics, are not a reliable source of food in many areas. In any event, game should not be shot unless necessary for survival. Berries and greens will add variety and vitamins to a survival diet, but few if any calories. Animals to be used for food should be thoroughly bled, internal organs removed, and the carcass chilled as soon as possible. This will prolong the keeping time of the meat. To expedite the chilling, clean snow can be packed in the body cavity. All meat should be cooked thoroughly as a safeguard against harmful microorganisms and parasites that might be present in the carcass. Only healthy animals should be used; in the absence of a person qualified to determine if the animal is healthy, meat from animals that appear sick should not be handled or eaten. For additional information, see FM 21-76.

51. Animals of Cold Regions

a. Caribou and Reindeer.

(1) These are mainly herd animals found in the high plateaus

and mountain slopes as well as in the grassy tundra areas. Their favorite year-round food is the lichens or "reindeer moss." Their summer diet consists of grasses, shrubs, and brush tips. They are very curious animals and will often approach a hunter merely from curiosity, thus presenting a good target. Sight of a human may have no effect on them but the slightest hint of human scent will send them galloping. It is possible to attract them near enough for a shot by waving a cloth and moving slowly toward them on all fours. In shooting, the aim should be for the shoulder or neck rather than the head.

(2) Reindeer have long been domesticated in Scandinavia and northern Asia for their meat, milk, hide, and as draft-animals.

(3) Both caribou and reindeer should be skinned promptly, especially in summer. Animal heat is the largest factor in meat spoilage. Fast and complete field dressing will eliminate most of this hazard and airing will finish the work. The bones and muscles can hold heat for as long as 48 hours, so the cooling process is very slow. Another hazard is the blowfly, which will ruin meat by laying eggs on it that turn into maggots in a few hours. Wiping off and drying the carcass will help keep the flies off. Fat should be kept with the carcass, not with the skin. If time does not allow skinning, at least the entrails and genitals should be removed and any excreta that may be present should be cleaned out of the animal. Pepper, if available, may be sprinkled on the carcass to provide a cover through which the blowflies will not work. An aerosol bomb may be used for spraying the brush in the immediate vicinity to keep flies out of the area.

(4) A poncho or insect bar may be used for wrapping the meat, whether for packing it out or if it is to be left hanging for the second trip. Meat should be raised off the ground as soon as possible because this will cool it sooner and keep it away from predators. Water will spoil meat quite rapidly. A carcass should never be washed until it has cooled and is ready to be butchered and stored. Rain will also hasten spoilage. The meat should be laid on piled up brush and covered with the hide until the rain stops.

b. Moose.

(1) The moose is the largest known species of the deer family. They are found in most areas of the northern hemisphere. Full grown bulls weigh from 1,000 to 1,200 pounds

and may stand six feet high at the shoulder. They require a large amount of forage and may usually be found in areas where food of this type is plentiful, such as in burn-offs, swamps, and lake areas. They forage on aspen, birch, and willow leaves, wild grass, and water plants. In the winter they can be spotted by climbing a hill or tree and looking for their "smoke" (ice fog).

(2) The procedures for the skinning and care of caribou and reindeer meat are applicable to moose.

c. *Seals.*

(1) Seals are widely distributed and generally common. Their flesh and liver are excellent food.

(2) In summer, the seals should be shot as they come to the surface of the water to breathe or as they are basking on rocks. The aim should be for the head. During the summer months when great quantities of fresh melt water lies on top of the salt, sometimes to a depth of as much as 6 feet, seals will generally sink to the level of the salt water and can be recovered with a pike pole or gang hook. In all other seasons of the year, most of the seals shot through the head will float, while about half of those shot through the body will not. In winter, seals will be found in the open leads in the ice pack or may be found at their breathing holes in the ice. However, hunting seals through breathing holes requires extreme patience and the holes are difficult to locate without the use of dogs.

(3) In the spring, mother seals and their pups may sometimes be located under snow hummocks adjacent to and over breathing holes, where they give birth to their young. In the spring, also, seals lie on the ice and bask in the sun. They must be carefully stalked and the hunter must be close enough at the time he shoots to dash and grab the dead seal before it slips down the incline and into the hole on the ice made slippery with blood.

(4) It takes great skill to stalk a seal. The Eskimo usually tries to imitate noises made by the seal, and he may use a white screen behind which he crawls while the seal sleeps, remaining absolutely still when the seal raises its head to look around. Seals normally sleep only for a few seconds at a time and then look around for their enemies for a few seconds before sleeping again. Seal meat from which the blubber has not been entirely removed will turn rancid in a short time.

d. Walrus. The meat and blubber of walrus are edible, as are the clams which may be found in their stomachs.

e. Bears. All bears are edible, although the flesh must be thoroughly cooked to guard against trichinosis. All bears are dangerous and hard to kill. There should be two or more hunters in the party when hunting bears; soft-nosed bullets should be used. The shoulder shot is best. If the bear stands up, the aim should be at the base of the throat and at center throat for a shot which will sever the vertebrae.

f. Wolves and Foxes. Wolves and foxes are edible. Wolves follow caribou herds. Arctic foxes follow polar bear and eat their leavings. Foxes will hang around a camp or follow a trail party and try to steal food.

g. Rabbits or Hares. Rabbits or hares can be snared or shot. They should be shot in the head or very little meat will be left. A whistle will probably cause a running one to stop long enough for an aimed shot. When cooking hare or rabbit, fat of some sort, should be added as the meat is very lean. They should not be dressed or cut up with bare hands because of the danger of contracting tularemia (rabbit fever) from contact with the raw flesh. Cooked flesh is safe to handle and eat.

h. Marmots. Marmots are woodchuck-like animals that live above the tree line in the mountains. They are excellent food, especially in late summer when they are very fat. The hunter should wait until the marmot moves away from his den before shooting or he may fall into his burrow.

i. Porcupines, Beavers, and Muskrats. These animals are found throughout the colder regions. Porcupines are excellent food, as are both beaver and muskrat. All are easily obtained. The porcupine, beaver, and muskrat can be easily killed with sticks. These animals move slowly on land.

j. Ground Squirrels. Ground squirrels abound in most cold areas and are easy to catch. They can be easily dug out of their burrows. They are especially common along streams with sandy banks.

52. Birds

All birds found in cold regions are edible. The eggs are also edible. For about 2 weeks during the summer molting season, most waterfowl lose their flight feathers and cannot fly. At this time they are easy targets and some can even be caught by hand. Certain nonmigratory birds are found in cold regions in wintertime. Several species of grouse, like the ruffed, sharp tail, spruce, and ptarmigan (which turn white in winter) are common. To obtain the greatest food value from birds, they should be plucked rather than skinned.

53. Fish

a. Fish form a large part of the native diet in cold regions and are almost the entire diet of work dogs in these areas. Along the coast, salmon, tomcod, flounder, sculpin, sand sharks, herring and other fish are found. Inland waters yield salmon, several varieties of trout, grayling, pike, candle fish, shee fish, ling cod, several varieties of whitefish, blackfish, and suckers. All fish and shellfish are edible, with the exception of the black mussel. Mussels from Pacific waters should be avoided entirely. Mussels are easily distinguished from clams and oysters by their orange-pink flesh. Shellfish can be cooked by throwing them into boiling water. Edible seafood is shown in figure 27.

b. Only the egg mass of the sea urchin is edible. To obtain eggs (which can be eaten raw) the sea urchin is cut open and the eggs scooped out. To prepare sea cucumbers, the insides are removed and the surface of the flesh scraped with a knife to remove the slimy skin. The flesh should be par boiled and then chopped up. They can then be cooked in a stew or fried. They require considerable cooking to make them tender.

54. Plants

In forested areas, food plants are most abundant along streams, sea shores, and in clearings. On the tundra, they are largest and most plentiful in wet places and in depressed areas protected from the wind. Many plants have starch stored in their roots. Small,

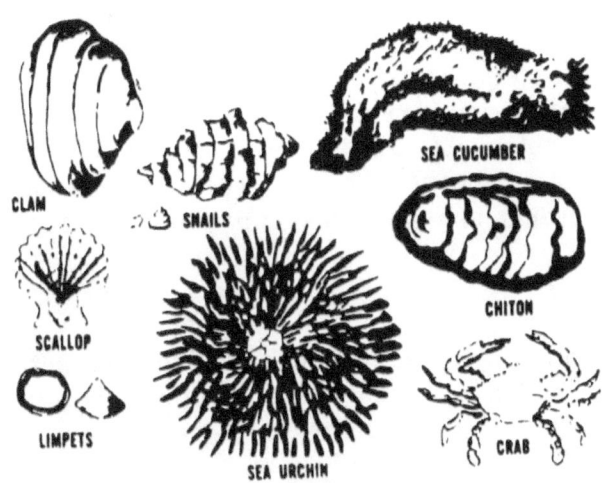

Figure 27. Edible seafood.

low-growing bushes often have many berries hidden under them. Edible parts of wild plants are often hidden. Edible food which might otherwise be overlooked, may be discovered by watching the feeding habits of animals, particularly birds. Reindeer or caribou moss is edible although it should be boiled, or at least soaked overnight, before eating.

a. All northern berries are safe to eat except the baneberry (fig. 28), which grows in southeastern Alaska and western Canada. Berry bushes are easily recognized in the forest, but on the tundra it is easy to overlook them because they are dwarfed. Many grow flat on the ground and are partly covered by mosses and lichens. A small bush may supply a handful of berries. In autumn, the leaves of certain berry bushes turn brilliant red or yellow, and show up as bright spots in mountain meadows or on the tundra.

b. On the barren tundra it is safe to eat any plant except one type of mushroom, the emetic russula. It can be recognized by the color of its "umbrella," which is pink or rosy when young, and red or yellowish when older.

c. In forested areas, all plants are edible except some mushrooms, water hemlock, and baneberry (fig. 28). Mushrooms have no significant caloric value, and since there is always a chance of picking a poisonous variety, it is recommended that mushrooms be eaten only when other foods are not available. All puffballs are edible but very young stages of some poisonous gill-bearing fungi may be mistaken for puffballs. If, when broken, the ball discloses gills, it should not be eaten. In forested areas, two fungi are especially poisonous: (1) death-cup amanita; and (2) the fly agaric (fig. 29).

d. The amanita is usually completely white but the cap may have tints of olive, purple, or brown. When fully grown the cap is 4 to 6 inches wide. The white gills on the underside of the cap are not attached to the stem. The stalk is white and brittle and has spherical base that is buried beneath the ground, resting in a soft white cup which is not visible unless the entire plant is exposed. The whole plant is poisonous. The fly agaric is a handsome fungus with yellow-orange or red-mottled cap, whitish or yellowish scales and white gills.

55. Water

Water points, operated by Corps of Engineer personnel, offer the best source of water supply for all troop units in any area and in any season. Under normal operating conditions, an Engineer unit with a water point capability will be attached to task forces of reinforced battle group size or larger. Engineer waterpoint

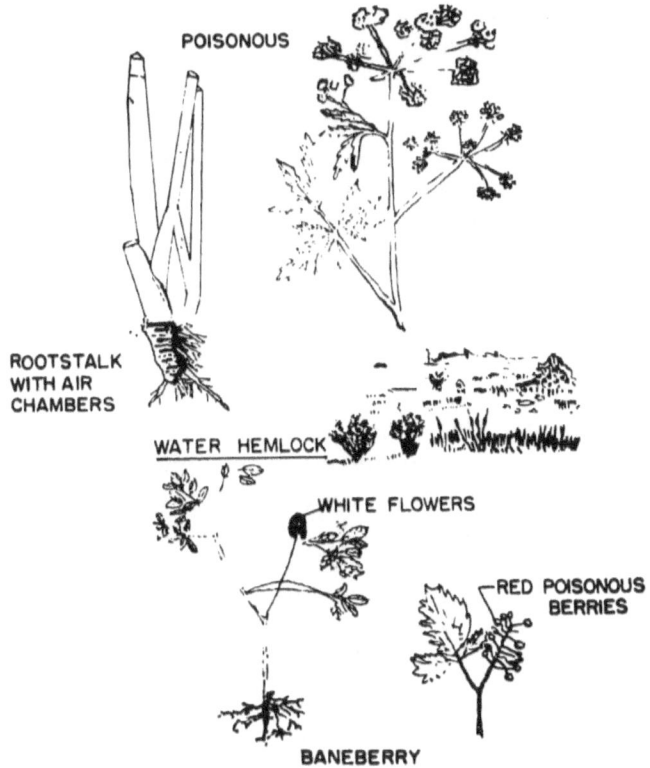

Figure 28. Water hemlock and baneberry.

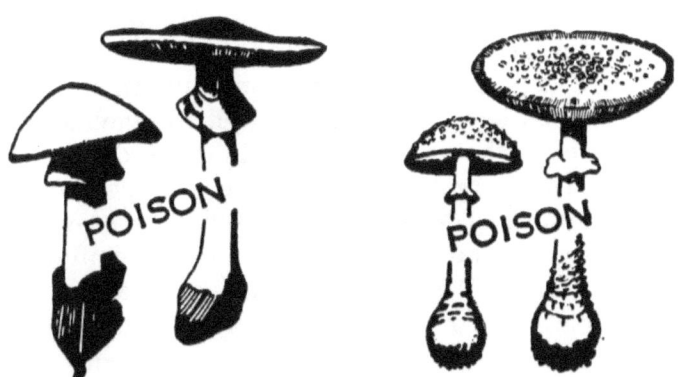

Figure 29. Deathcup amanita and fly agaric.

operations under cold weather conditions are discussed in FM 31–71. This paragraph, together with paragraphs 56 and 82 offers possible solutions to the problem of water supply that confronts individuals and small detachments operating in isolated areas away from normal support activities.

 a. Water is plentiful in most cold regions in one form or another. Potential sources are streams, lakes and ponds, glaciers, fresh-water ice, and last year's sea ice. Freshly frozen sea ice is salty, but year-old sea ice has had the salt leached out. It is well to test freshly frozen ice when looking for water. In some areas, where tidal action and currents are small, there is a layer of fresh water lying on top of the ice; the lower layers still contain salt. In some cases, this layer of fresh water may be 2 to 4 feet in depth.

b. If possible, water should be obtained from running streams or lakes rather than by melting ice or snow. Melting ice or snow to obtain water is a slow process and consumes large quantities of fuel. Seventeen cubic inches of uncompacted snow, when melted, yields only 1 cubic inch of water. In winter a hole may be cut through the ice of a stream or lake to get water; the hole is then covered with snow blocks or a poncho, board, or ration box placed over it. Loose snow is piled on top to provide insulation and prevent refreezing. In extremely cold weather, the waterhole should be broken open at frequent intervals. Waterholes should be marked with a stick or other marker which will not be covered by drifting snow. Water is abundant during the summer in lakes, ponds, or rivers. The milky water of a glacial stream is not harmful. It should stand in a container until the coarser sediment settles.

c. In winter or summer, water obtained from ponds, lakes, and streams must be purified by boiling or by treating with water purification tablets.

d. Under chemical, biological, and/or atomic warfare, precautions should be taken against using contaminated water sources. In general, cold weather conditions tend to prolong or conceal contamination hazards, and unexpected contamination may thus be encountered. When snow or ice is thawed to provide water supplies, detection tests should be conducted during or after the melting operation, since frozen contamination may not be detectable. Radiological contamination which has been covered with snow or ice may or may not show up on radiac instruments, depending upon the thickness of the cover. Boiling or treating with water purification tablets has no effect on radioactive contaminants in water. In emergencies, water suspected of radiological contamination may be filtered through a 6-inch column of loose dirt and then treated with water purification tablets. Purification of water showing, or suspected of containing, chemical contamination should not be attempted.

e. After the water is obtained, the problem of transporting and storing it arises. Units operating in the field under cold weather conditions may store water in insulated 5-gallon thermos jugs, 5-gallon water cans with insulated covers, or other similar type containers for use by small detachments or individuals. Immersion-

type heaters may be used to prevent freezing of water supply tanks. Some points to be remembered are:
  (1) Transportation of water by wheeled vehicles in barren, sparsely settled areas under snow and ice conditions is practicable only when there is a road net established. The best way to transport water in cold regions is by the use of track-laying vehicles which are not dependent on roads for maneuverability. If 5-gallon cans are used to carry water, they are filled only three quarters full to allow agitation of the water and help prevent freezing while in transit. Cans are stored off the floor in heated shelters as soon as they are delivered. Sled-mounted, 250 to 300-gallon water tanks in which immersion-type heaters have been installed have proved satisfactory.
  (2) For small units of two to four men, the 5-gallon insulated food container is satisfactory for water storage. These can be filled at night and will hold enough water for the next day's needs for about four men. The insulation of these containers is sufficient to keep water from freezing for as long as 40 hours at an ambient temperature of —20° F., if the temperature of the water was at boiling point when the container was filled.

56. Types of Ice and Snow

  a. When water is not available from other sources, it must be obtained by melting snow or ice. To conserve fuel, ice is preferable when available; if snow must be used, the most compact snow in the area should be obtained. Snow should be gathered only from areas that have not been contaminated by animals, humans, or toxic agents.

  b. Ice sources are frozen lakes, rivers, ponds, glaciers, icebergs, or old sea ice. Old sea ice is rounded where broken and is likely to be pitted and to have pools on it. Its underwater part has a bluish appearance. Fresh sea ice has a milky appearance and is angular in shape when broken. Water obtained by melting snow or ice may be purified by use of water purification tablets, providing it has not been contaminated by toxic agents.

  c. If chemical, biological, or radiological contamination is detected, procedures as outlined in paragraph 55*d* will be followed.

57. Procedures for Melting Snow and Ice

  a. Burning the bottom of a melting pot by "priming" can be avoided by placing a small quantity of water in the bottom of the pot and adding snow gradually. If water is not available, the pot should be held near the stove and a small quantity of snow melted in the bottom of the pot before filling it with snow.

b. The snow should be compacted in the melting pot and stirred occasionally to prevent burning the bottom of the pot.

c. Pots of snow or ice should be left on the stove when not being used for cooking so as to have water available when needed.

d. Snow or ice to be melted should be placed just outside the shelter and brought in as needed.

Section V. HYGIENE AND FIRST AID

58. General

In cold weather, the care of the body requires special emphasis. If men are allowed to go without washing, fail to eat properly, do not get sufficient liquids or salt, efficiency will suffer. Lowered efficiency increases the possibility of casualties, either by cold injury or enemy action.

59. Dehydration

a. Definition and Principle. Dehydration means to lose or be deprived of water or the elements of water. A growing plant loses (uses) water in the growing process. If this water is not replaced by either natural means (rain) or by watering, the plant will wither and eventually dry up. The same principle applies to the human body which loses water and, an additional element, salt. A certain amount of this loss is taking place constantly through the normal body processes of elimination; through the normal daily intake of food and liquids, these losses are replaced.

b. Dangers. When individuals are engaged in any strenuous exercises or activities, an excessive amount of water and salt is lost through perspiration. This excessive loss creates what is known as "imbalance of liquids" in the body and it is then that the danger of dehydration arises, unless this loss of liquids and salt is replaced immediately and individuals are allowed sufficient rest before continuing their activities.

c. Training and Discipline. The danger of dehydration for troops operating under cold weather conditions and over ice and deep snow is a problem that does exist and cannot be overemphasized. It is equally important, however, to recognize that the problem can be overcome and will present no great obstacle to well trained, disciplined troops who have been thoroughly oriented in the causes, the symptoms, and the effects of dehydration and who have been properly instructed in preventive measures.

d. Differences. It is important, therefore, to be aware that the danger of dehydration is as prevalent in cold regions as it is in hot, dry areas. The difference is that in *hot weather* the individual is conscious of the fact that the body is losing liquids and salt because he can see and feel the perspiration with its saline taste and "feel" it running down the face, getting in the eyes, and on the lips and tongue, and dripping from the body. In *cold weather*, it is extremely difficult for an individual who is bundled up in many layers of clothing to realize that this condition does exist. Under these conditions, perspiration is rapidly absorbed by the heavy clothing or evaporated by the air and is rarely visible on the skin.

e. Cause, Symptoms, Effects, Preventive Measures, and Treatment.

(1) Dehydration is caused by failure to correct the body's "imbalance of liquids" through replacing liquid and salt which has been lost, and by failure to allow sufficient rest when engaged in strenuous activities.

(2) The symptoms of dehydration parallel the symptoms of sunstroke and heat exhaustion. The mouth, tongue, and throat become parched and dry and swallowing becomes difficult. General nausea is felt and may be accompanied by spells of faintness, extreme dizziness and vomiting. A feeling of general tiredness and weakness sets in and body aches are felt, especially in the legs. It becomes difficult to keep the eyes in focus and fainting or "blacking out" may occur.

(3) The effect of dehydration on the individual is to incapacitate him for a period of from a few hours to several days. The effectiveness of the individual's unit is likewise reduced by the loss of his contribution to the accomplishment of the unit mission. Small patrols and detachments operating beyond range of immediate help from the parent unit must be extra-cautious to avoid dehydration since they run the risk of a secondary but more dangerous effect of dehydration, that of becoming cold weather casualties while incapacitated.

(4) Dehydration can be prevented during cold weather operations by following the same general preventive measures applicable to hot, dry areas. Salt and sufficient additional liquids are consumed to offset excessive body losses of these elements. The amount will vary according to the individual and the type of work he is doing, i. e., light, heavy, very strenuous, etc. Rest is equally important as a preventive measure. Each individual must realize that any work that must be done while bundled in several

layers of clothing is extremely exhausting. This is especially true of any movement by foot, regardless of how short the distance.

(5) In treating a person who has become dehydrated, the individual should be kept warm but his clothes loosened sufficiently to allow proper circulation; liquids and salt should be fed to him gradually and, most important of all, he must have plenty of rest. When salt tablets are not available, common table salt may be used. Approximately one-third of a level mess spoon of salt mixed with a canteen cup full of water makes a palatable solution.

60. Personal Hygiene

Because of the extremes in temperatures and lack of bathing and sanitary facilities, keeping the body clean in cold weather will not be easy.

(1) The entire body should be washed at least weekly. If bathing facilities are not available, the entire body can be washed with the equivalent of two canteen cups of water, using half for soaping and washing, and half for rinsing. If circumstances prevent use of water, a rubdown with a dry cloth will help. The feet, crotch, and armpits should be cleaned daily.

(2) A temporary steam bath can be built in a large-size tent (fig. 30). Cobblestones are piled up to form a furnace. The furnace is either heated inside the tent (ventilation flaps wide open) or in the open with the tent pitched over the furnace after the stones are heated. Wood is used for fuel. Seats and water buckets are taken into the tent after the stones are nearly red-hot and the fire has died down, so that they do not get sooty. The pouring and washing water is usually heated outside the tent. The water is thrown on the hot stones in small quantities. Thus it does not drop into the ashes and the temperature does not rise too fast. A naked person spends from 15 minutes to 1 hour in this steam bath. After thoroughly perspiring, the body is washed with tepid water.

(3) Beards should be shaved or clipped close. Hair should be combed daily and not allowed to grow too long. A beard or long hair adds very little in insulation value and soils clothing with the natural hair oils. In winter, a beard or a mustache is a nuisance since it serves as a base for the buildup of ice from moisture in the breath and will mask the presence of frostbite. All individuals should shave daily, when possible. Under chemical or biological

warfare conditions, a beardless face and daily shaving are especially important, since an airtight seal of the protective mask is difficult to obtain with even stubble on the face.

(4) Socks should be changed and the feet washed daily. If this is not possible, the boots and socks should be removed, and the feet massaged and dried. By sprinkling the feet liberally with foot powder and then rubbing the powder off, the feet can be efficiently dry cleaned.

(5) Sleeping bags should be kept clean. Liners should be used if issued. Subject to operational requirements, the best method is to wear only the minimum clothing in the bag and never get into it with damp underwear. Dry underwear should be put on before going to sleep and the other set hung up to dry. Perspiration will soil a sleeping bag, and cause it to become damp; therefore, the bag should be aired as frequently as possible. In the morning, the bag should be opened wide and air pumped in and out to remove the moist air within the bag.

(6) Teeth should be cleaned daily. If a toothbrush is not available, a clean piece of gauze or other cloth wrapped around the finger, or end of a twig chewed into a pulp may be used in lieu of a toothbrush.

(7) Underwear and shirts should be changed at least twice weekly; however, if it is not possible to wash the clothing this often the clothing should be crumpled, shaken out, and aired for about 2 hours.

61. Health Hazards and First Aid

Cold injuries, which include frostbite, trenchfoot, or general chilling of the body, are all caused by lack of proper insulation against the cold.

a. Frostbite is the freezing of some part of the body by exposure to temperatures of freezing or below. It is a constant

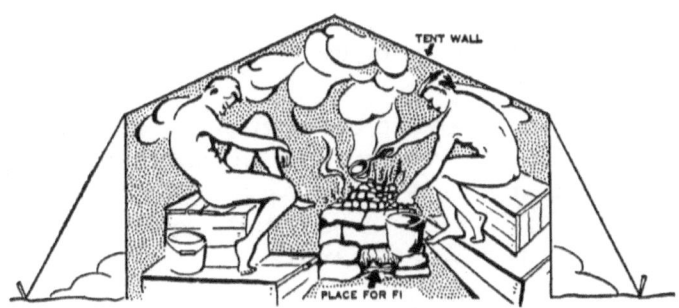

Figure 30. Temporary steam bath in tent.

hazard in subzero operations, especially when the wind is strong. Usually there is an uncomfortable sensation of coldness followed by numbness. There may be a tingling, stinging, or aching sensation, even a cramping pain. The skin first turns red. Later it becomes pale or waxy white. Depending on the damage caused to the tissue, the following classification is given to the frostbite:

1st degree: Mild—serves as warning—no problem.

2d degree: Outer skin blisters, but not the flesh beneath — painful, may require medical aid to relieve (fig. 31).

3d degree: Skin dies, flesh under it dies, loss of limb possible.

(1) It is easier to prevent frostbite, or stop it in its very early stages, than to thaw and take care of badly frozen flesh. Clothing and equipment must be fitted and worn so as to avoid interference with circulation. To prevent severe frostbite:

   (*a*) Sufficient clothing must be worn for protection against cold and wind. Gloves or socks with holes in them should not be worn. The face must be protected in high wind, and when exposed to aircraft prop blast.

   (*b*) Any interference with the circulation of the blood reduces the amount of heat delivered to the extremities. All clothing and equipment must be fitted and worn to avoid interference with the circulation.

   (*c*) Every effort must be made to keep the clothing and the body as dry as possible.

   (*d*) Cold metal should not be touched with the bare skin in extreme low temperatures. Adequate clothing and shelter must be provided especially during periods of inactivity.

   (*e*) The face, fingers, and toes should be exercised from time to time to keep them warm and to detect any numb or hard areas. The ears should be massaged from time to time with the hands for the same purpose.

   (*f*) The buddy system should always be used. Men should pair off and watch each other closely for signs of frostbite. Any frozen spots must be thawed immediately, using bare hands or other sources of body heat.

(2) In case of frostbite:

   (*a*) The casualty should be taken to heated shelter if possible. All constricting items of clothing such as boots, gloves, and socks should be removed from the area of injury if it can be done without causing further damage to the frostbitten part.

   (*b*) The frozen part should be placed against an unfrozen portion of the body or exposed to warm air. The general body warmth should be maintained by wrapping the patient in blankets or similar material. Interference with circulation must be avoided when bandaging the injured part or when preparing the casualty for evacuation. To protect the injured area, the cleanest material available should be used for bandages; sterile dressing should be used if available. Medication or ointment of any kind should not be put on the injured area. Premedicated bandages such as petrolatum-gauze should not be used. Warm food and drink will help restore normal body temperatures. The warm drink should be nonalcoholic and the individual should not be allowed to smoke, as both alcohol and tobacco have an adverse effect upon proper circulation of the blood. In severe cases of prolonged exposure it may be necessary to use artificial respiration during the early phases of treatment.

(c) Every attempt should be made to employ body heat to aid in thawing. A bare, warm palm may be held against frostbitten ears or parts of the face. A frostbitten wrist should be grasped with a warm, bare hand. Frostbitten hands should be held against the chest, under the armpits, or between the legs at the groin. A frostbitten foot can be held against a companion's stomach or between his thighs.

(d) Frostbite should never be rubbed. This may tear frozen tissues and cause further tissue damage. Snow or ice should never be applied to the area; cold injury is increased by so doing. For the same reason, frozen limbs should never be soaked in kerosene or oil.

(e) No attempt should be made to thaw frozen parts by exercising them or by exposing them to open fire. Such measures will increase tissue damage and are likely to break the skin.

(f) Litter or sled evacuation should be used in cases of frozen lower extremities, to prevent further injury. The casualty should be placed in a casualty evacuation bag, if available, or a sleeping bag, and evacuated to a heated shelter immediately.

b. Trenchfoot is a cold injury resulting from prolonged exposure to temperatures near freezing. It is important to remember that the temperature does not need to be below 32° F. to cause this injury. In the early stages of trenchfoot, feet and toes are pale and feel cold, numb, and stiff. Walking becomes difficult. If preventive action is not taken at this stage, the feet will swell

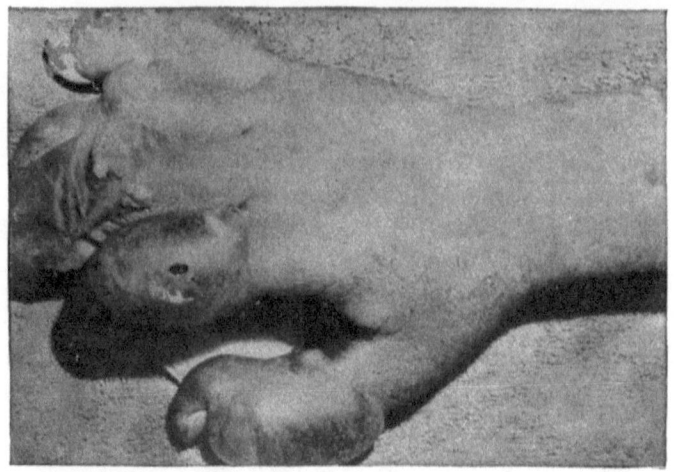

Figure 31. Frostbitten hand (2d degree frostbite).

and become painful. In extreme cases of trenchfoot the flesh dies and amputation of the foot or of the leg may be necessary. Because the early stages are not painful, individuals must be constantly alert to prevent the development of trenchfoot. To prevent this condition:

(1) Feet should be kept dry by wearing waterproof footgear and by keeping the floor of shelters dry.

(2) Socks and boots should be cleaned and dried at every opportunity.

(3) The feet should be dried as soon as possible after getting them wet. They may be warmed with the hands. Foot powder should be applied and dry socks put on.

(4) If it becomes necessary to wear wet boots and socks, the feet should be exercised continually by wriggling the toes and bending the ankles. Tight boots should never be worn.

(5) In treating trenchfoot, the feet should be handled very gently. They should not be rubbed or massaged. If necessary, they may be cleansed carefully with plain white soap and water, dried, elevated, and allowed to remain exposed. While it is desirable to warm the patient, the feet should always be kept cool by exposure to room air. The casualty should be removed to a medical facility as soon as possible. He should be carried and not permitted to walk on damaged feet.

c. Severe chilling results from total immersion in cold water for even a few minutes. If this occurs the body temperature will drop. Longer exposure to severe dry cold can also lower the body temperature. The only remedy for this severe chilling is warming of the entire body. The casualty must be warmed by any means available. Severe chilling is often accompanied by shock, and the victim should be evacuated to an aid station as soon as possible.

d. The length of time that a casualty may be exposed to the weather without danger of cold injury due to exposure varies directly with the temperature level and the wind velocity. The lower the temperature and the stronger the wind, the sooner does chilling occur. There is a great variation in individual reactions to cold. To give competent care to the injured in extreme cold, the medical personnel must have heated shelter in which to operate. Battle wounds in the cold are no different from those sustained in more temperate climates, and should be treated in the same manner. Morale is helped by the assurance that the sick and wounded can be rapidly transported from the battlefield to hospitals, and that for the non-transportable cases requiring prompt life-saving surgery, hospitals with highly skilled surgical personnel are available at clearing station level.

62. Shock

Shock is a condition characterized by a reduction in the effective circulating blood volume. This can be brought about by severe injuries, loss of blood, pain, emotional disturbances, or any of many other factors. The normal reaction of the body to severe cold, reduction of the volume of blood circulating to extremities, is very similar to the reaction of the circulatory system to the condition of shock. Shock will usually develop more rapidly and progress more deeply in extreme cold than in normal temperature.

a. Signs of Shock. The signs of shock are apprehension; sweating; pallor; rapid, faint pulse; cold clammy skin; and thirst. If the patient is not given good first aid treatment immediately the condition of shock may progress until the patient passes into unconsciousness and further into death.

b. First Aid for Shock.
  (1) The injured person should be made as comfortable as possible.
  (2) Pain may be relieved by proper positioning, good bandaging and splinting. Aspirin will also help, if it is available.
  (3) The litter should be positioned so that the casualty's head is lower than his trunk and legs. This should not be done if it will cause discomfort to the patient.
  (4) The patient should be kept warm with blankets and sleeping bags.
  (5) When the patient is conscious he should be given warm soup, chocolate, coffee, or tea.

(6) The patient should receive medical attention as soon as possible.

63. Sunburn

An individual may get sunburned when the temperature of the air is below freezing. On snow, ice, and water, the sun's rays reflect from all angles; in a valley the rays come from every direction. Sunlight reflected upward from the bright surfaces attacks men where the skin is very sensitive—around the lips, nostrils, and eyelids. Sunburn cream and a chapstick should be carried in the pocket, and applied to those parts of the face that are exposed to direct or reflected light. Soap or shaving lotions with a high alcoholic content should not be used because they remove natural oils that protect the skin from the sun. If blistered, report to an aid station as soon as possible, as the blistered area, especially lips, may become badly infected.

64. Snow Blindness

Snow blindness occurs when the sun is shining brightly on an expanse of snow, and is due to the reflection of ultraviolet rays. It is particularly likely to occur after a fall of new snow, even when the rays of the sun are partially obscured by a light mist or fog. In most cases, snow blindness is due to negligence or failure on the part of the soldier to use his sunglasses. Symptoms of snow blindness are a sensation of grit in the eyes, pain in and over the eyes, watering, redness, headache, and photophobia (distaste for light). A poultice of cold, used tea leaves may be used to give relief if no medications are available. Dark eye shades or bandages should be placed over the eyes.

65. Constipation

a. When operating under cold weather conditions there is a general tendency for individuals to allow themselves to become constipated. This condition is brought about by the desire to avoid the inconvenience and discomfort of relieving themselves under adverse conditions. This condition is also caused by changes in eating habits and failure to drink a sufficient amount of liquids. If this condition exists for a prolonged period, it could result in mild to severe stomach cramps, headaches, dizziness, and the setting-in of a feeling of general tiredness and weakness. An individual so affected may become incapacitated for some time.

b. Constipation can usually be prevented by adjusting the normal eating and drinking habits to fit the activities in which engaged, and by not "putting off" the normal, natural, processes of relieving the body of waste matter. Medical personnel should be consulted if severe constipation persists. Each individual must be educated concerning the consequences of neglecting personal hygiene habits.

66. Carbon Monoxide Poisoning

a. Whenever a stove, fire, or gasoline heater is used indoors, there is danger of carbon monoxide poisoning. A steady supply of fresh air in living and working quarters is vital. Carbon monoxide is a deadly gas, even in low concentration, and is particularly dangerous because it is odorless.

b. Generally there are no symptoms. With mild poisoning, however, these signs may be present—headache, dizziness, yawning, weariness, nausea, and ringing in the ears. Later on, the heart begins to flutter or throb. But the gas may hit without any warning whatsoever. A soldier may not know anything is wrong until his knees buckle. When this happens, he may not be able to walk or crawl. Unconsciousness follows; then death. Men may be fatally poisoned as they sleep.

c. In a case of carbon monoxide poisoning, the victim must be moved into the fresh air at once, but must be kept warm. In the winter, fresh air means merely circulating air that is free from gases. Exposure to outdoor cold might cause collapse. If the only fresh air is outdoors, the patient should be put into a sleeping bag for warmth. A carbon monoxide victim should never be exercised, because this will further increase his requirements for oxygen. If a gassed person stops breathing or breathes only in gasps, artificial respiration should be started immediately. In the latter case, the operator's movements must be carefully synchronized with the victim's gasps. Breathing pure oxygen removes carbon monoxide from the blood faster than does breathing air and greatly hastens recovery. Carbon monoxide poisoning is serious and a victim who survives it must be kept absolutely quiet and warm for at least a day. Hot water bottles and hot pads are helpful in maintaining body temperatures.

67. Care of Casualties

If any member of a group is injured, the most important course of action is to get him to competent medical aid as soon as possible. The casualty should be given first aid treatment, protected from the cold and shock effects, and evacuated to an aid station with a minimum of delay. He should be placed in a casualty bag, sleeping bag, or the best available substitute. He should have warm drinking water or other hot drinks, except in the case of abdominal injury.

Warning: **Once a tourniquet has been applied, the wounded man should be examined by a medical officer as soon as possible.**
If possible, the tourniquet should not be loosened by anyone except a medical officer who is prepared to stop the hemorrhage or bleeding by other means and to administer other treatment as necessary. Repeated loosening of the tourniquet by inexperienced personnel is extremely dangerous, can result in considerable loss of blood, and endanger the life of the patient. Halting of circulation to the extremities is an invitation to frostbite. If morphine is to be administered, caution must be exercised to avoid overdosage. A circulation constricted by cold exposure absorbs morphine slowly.

68. Emergency Evacuation

Personnel who have been wounded should be evacuated to the nearest medical facility by the fastest means of transport available. Sleds can be used if oversnow vehicles or air evacuation facilities cannot be obtained. It may be necessary to use man-hauled sleds to move the wounded a safe distance behind the front lines before they can be transferred to faster means of transport (fig. 32). Speed in evacuation is essential because of the combined effects of severe cold and shock on the wounded.

69. Body Parasites

a. General. Body parasites are very common in the more populated cold regions because of the crowded living conditions and shortage of bathing and cleaning facilities. When in the midst of a native population, or when occupying shelters which have been used before, individuals must inspect clothing and body each night for body parasites.

Figure 32. Evacuation of wounded on sled.

b. Means of Control. If clothing has become infested with lice, the following methods of removing them are recommended:

    (1) While extreme cold does not kill lice, it paralyzes them. The garments should be hung in the cold; then beat and brushed. This will help rid the garments of lice, but not of louse eggs.

    (2) DDT powder and other chemical means can be used to free the body and clothing of body parasites.

70. Insects

a. Mosquitoes. Mosquitoes are abundant because of the numerous swamps and lakes, and the long, sunny days with slight variation in heat from night to day provide ideal conditions for insect incubation. Both malaria and yellow fever, transmitted by mosquitoes in other parts of the world, are unknown in the northern regions.

b. Midges. These extremely small bloodsucking insects, also known as "punkies," "no-see-ums," and "creeping fire," are found in many areas of the North in such numbers that they become terrible pests. Midges are most abundant in middle and late summer; they breed in decaying leaves, along stream margins, or even in holes in trees. They bite chiefly on cloudy days, in the evening and very early in the morning; in bright daylight they seek the shade. Their most troublesome aspect is their small size. They easily penetrate even the finest mosquito netting and are excluded only by 60-mesh silk bolting cloth.

c. Blackflies. Small, chunky, blackish gnats, called blackflies or buffalo gnats, swarm about in large numbers in the early northern summer, chiefly in forested areas. They breed in swiftly flowing streams and hover about the eyes, ears, and nostrils, making little noise but promptly alighting and sucking blood wherever skin is exposed, especially behind the ears. While their bites are not especially painful, sensitive persons are affected by itching and swelling from the bites. Blackflies are usually active only in daylight and generally are most numerous during June and July.

d. Biting Flies. In the northern areas there are three kinds of large flies which are—deerflies, mooseflies, and horseflies. All these "Biting Flies," like the blackflies, frequent marshes and bogs on hot, bright days. Strong winds do not keep them down, but cool weather sends them to cover. They are extremely annoying to man because of their continual circling and buzzing, as well as their vicious bites. Under certain conditions, deerflies are capable of transmitting tularemia.

e. Preventive Measures.

(1) *General.* Protective covering for the entire body is the best method of combating insect pests. Repellents in the form of oils and ointments are invaluable against some kinds of insects. In camp, smudges can be used but are of doubtful value. Insect-proof shelters are often a necessity.

(2) *Clothing.* The complete uniform should be worn so that insects will be unable to reach the skin. The cloth should be so thick or so closely woven that insects cannot penetrate it. Zippered or pullover shirt fronts are preferred to open fronts. To prevent flies from attacking ankles, high shoes should be worn or trouser bottoms wrapped around the ankles and tied with strips of cloth. Gloves are necessary and must be long enough to close the wrist openings. If available, the issue head net or sunglasses with screen sides should be worn. If not, insect repellent should be used. Insect control measures in camp are covered in paragraph 85.

71. Safety

a. Fire is a constant hazard under winter conditions because of the great number of tents and stoves used. Whenever a stove or open fire is used, it should never be left unguarded. The fire prevention and safety measures in care and operation of tent stoves, candles, and lanterns must be strictly enforced.

b. Heavy equipment and tools are a source of danger. Great care must be exercised by the individuals using tools while wearing gloves. Ice or water on gloves may cause the tool to slip from the user's hand, causing injury.

c. Terrain hazards such as steep rock faces concealed by snow, glacier travel, and crevasses and avalanches are typical, especially in mountain operations. Unfrozen rivers with strong currents and lakes with thin ice are also constant hazards in northern operations. Guides familiar with terrain peculiarities must be used to the greatest extent during the troop movement.

Section VI. BIVOUAC ROUTINE

72. Location of Bivouac Sites

The selection of bivouac sites in northern areas is all-important and requires careful consideration. The problem of selection varies with the tactical situation, terrain, and weather conditions.

a. If possible, the bivouac area should be tactically located in accordance with the principles of security and defense. It should be located so that it would be advantageous for future operations. If contact with the enemy is imminent, the bivouac should be located on high ground; this, at times, is disregarded in favor of cover and concealment, more suitable ground conditions, etc.

b. Cover and concealment against air and ground observation is essential for the bivouac area. Forested areas pose few problems in comparison to that area north of the tree line. Particular attention must be given in selecting areas in cold regions to insure that local camouflage materials are available in open areas.

c. In the winter, protection from the wind is a prime consideration. This is particularly true in the Far North, where violent local gales occur frequently. In wooded areas the wind has little effect. In the summer a windy area will generally be free of insects and it will also be helpful in dispersing smoke from the bivouac area.

d. The condition of the ground is important and, if possible, the bivouac should be located on hard, dry ground.

e. Construction materials play an important part in the selection of a bivouac. When making a reconnaissance for the area, such things as the availability of firewood, water, snow for snow shelters, boughs, etc., must be considered.

73. Bivouac in Forests

a. Most forests in cold regions provide excellent bivouac sites and should be utilized whenever possible. Forests provide many natural materials such as boughs for insulation, firewood, and camouflage construction materials. They also provide excellent cover and concealment against enemy air and ground observation. Coniferous forests (cone-bearing trees) provide better protection from wind and better insulation material and firewood than deciduous forests. Pine and spruce forests, normally found on well drained soil, offer the best hardstand for shelter.

b. Tracks are visible in both summer and winter. On dry ground, however, they normally are not as noticeable as on wet soil. Consideration should be given to building dummy positions for the purpose of misleading the enemy (fig. 33). Track discipline must be rigidly enforced in the bivouac area. Once tracks are made, all movement within the area should be restricted to those tracks.

74. Bivouac on Marshy Ground

a. In winter, when the ground is frozen, good bivouac sites oftentimes may be found in wet areas if local concealment is good. In summer, swampy areas provide poor facilities for the

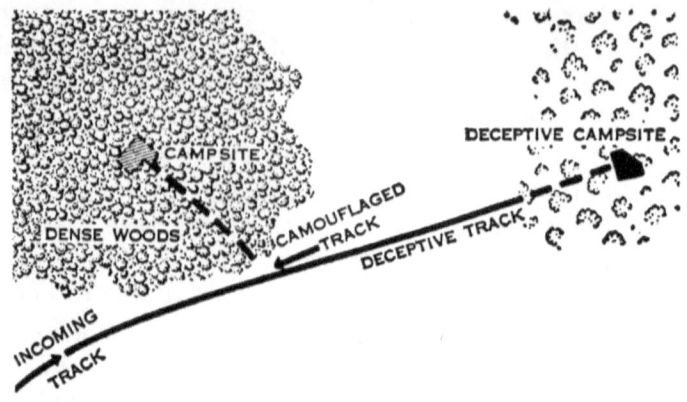

Figure 33. Selection of route when entering bivouac in forest.

bivouac site. If it becomes mandatory to establish the bivouac on swampy ground, flooring for the shelters must be constructed. If tree trunks are available, a "float" may be built under the shelter (fig. 34). In the absence of the tree trunks, brush matting will serve the same purpose.

b. Areas to be used for extended periods of time require draining, clearing of existing creeks, digging of ditches around the shelter, or preparing a water trench inside the shelter.

c. Swarms of insects frequently are bothersome in the summer, so it is desirable to locate the bivouac in areas where there is wind. Such areas can be found along a river bank or on a lake with an onshore breeze. Smoke smudges are sometimes helpful in driving insects away from the area. Normally, river banks and shores of lakes provide the best vegetation, the materials needed for the bivouac, and water and concealment.

75. Bivouac in Open Terrain and on Ice

a. Due to strong winds, drifting snow, and poor concealment, bivouac areas in the barren tundra must be chosen carefully.

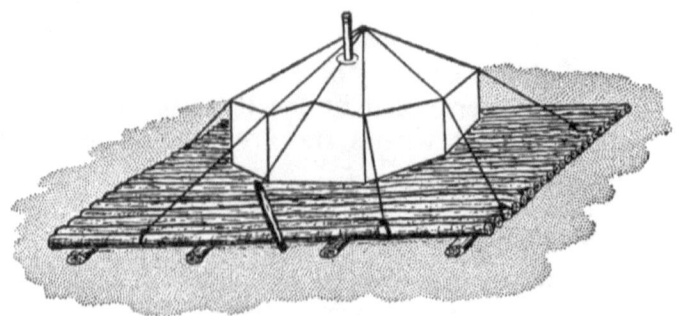

Figure 34. "Float" under shelter.

b. Tents should be pitched where they can be sheltered by natural windbreaks whenever possible. The windbreak may consist of depressions in the ground or pressure ridges on the ice. A visual inspection will indicate the degree of drifting, direction of the prevailing wind, and more suitable protected areas for locating the shelters. In areas where natural windfalls do not exist, snow walls (fig. 35) may be constructed to provide protection from winds and enemy small arms fire, as well as concealment from ground observation. In open areas with high winds, snow gathers rapidly on the lee side, making it necessary to clear the sides and tops of the tents periodically to prevent the weight of the drifting snow from collapsing the tent. The entrance to the shelter should face downwind from the prevailing wind. This will prevent the snow from blocking the exit and cutting off the ventilation.

Figure 35. A tent dug in with snow wall windbreak.

c. When the tent is pitched on ice, holes are chopped where the tent pins are normally set. "Deadmen" are inserted in the holes at right angles to the tent. The holes are then packed with snow or filled with water and left to freeze.

76. Bivouacs in Mountains

a. Mountainous terrain is characterized by strong winds, cold and lack of concealment above the timberline. The wind overhead creates an extensive lee near the mountain. The overhead lee resembles the dry space behind waterfalls caused by water having such speed that it shoots over the edge of the cliff and descends in a curve. An inland wind blowing 50 miles an hour may not strike the ground for several miles after passing the edge of a cliff or a very steep slope.

b. Cold air is heavier and frequently settles in valleys. The point where the temperature starts changing is low in summer and higher and more noticeable in winter. Therefore, in some instances it is better to establish a bivouac up the hillside above the valley floor and below the timberline, where applicable. Avalanche hazard areas must be carefully avoided.

77. Establishing Bivouac

a. General. Setting up a bivouac is a routine based on specific drills and procedures which enable the commander to control the bivouac area, have it always protected, camouflaged, and the personnel ready to fight. Only the minimum amount of time should be devoted to pitching and striking the shelters and to general housekeeping. Bivouacking in a routine manner allows more time for daily movement, establishing an effective security system, and defense of the bivouac site. Finally, it allows more time for rest and to make preparations for the continuation of the operation.

b. Responsibilities of Unit Leader. On entering the bivouac site, the unit leader is responsible for:

(1) Posting a security guard.
(2) Checking the bivouac site.
(3) Determining exact tent locations providing the best natural shelter and camouflage.
(4) Designating an area from which construction material and firewood will be obtained.
(5) Selection of a water point, or marking off the snow area to be utilized for water.
(6) Designating latrine and garbage disposal sites.
(7) Designating a site for weapon and ski racks. Temporary placement for weapons and equipment must be arranged until the bivouac has been established.
(8) Breaking a minimum number of trails between the tent site and area assigned for firewood and construction material, water point, and latrine.
(9) Maintaining camouflage and track discipline at all times.
(10) Organization and assignments for the work details as follows:
   (*a*) Clearing and leveling the shelter sites. In winter the snow is dug to the ground level or, in emergency, packed down by trampling with skis or snowshoes.
   (*b*) Pitching tents (when used).
   (*c*) Cutting, trimming, and hauling trees and boughs for construction of improvised shelters and bough beds (when tents are not available).
   (*d*) Construction of improvised shelters best suited to the area concerned.
   (*e*) Construction of windbreaks, if necessary.
   (*f*) Building necessary weapon and ski racks. Special care must be given to the protection of the weapons from the elements.

(g) Construction of field latrines and garbage disposal sites.

(h) Preparing a water point.

(i) Gathering and cutting a supply of firewood.

(j) During cold weather, situation permitting, starting fires and preparing hot drinks for all individuals.

(k) Upon completion of shelter construction, starting warm meal.

(11) Maintaining and emphasizing cleanliness, tidiness, and teamwork.

(12) Upon completion of the bivouac, arranging equipment within and outside of shelters.

(13) Preparing defensive positions and breaking and marking a trail from the shelters to the positions.

(14) Maintaining a duty roster for exterior guards, fire guards, and similar assignments.

(15) Rotating individuals on all jobs on a daily basis.

(16) Assigning specific sleeping areas for all individuals in accordance with the duty roster.

(17) Upon establishing the bivouac, removing the exterior guard in case the parent unit has taken over the security of the area.

(18) Inspecting the area, examining the security, camouflage, cover, weapons, skis, sleds, vehicles (if applicable), and the conditions of the men and their equipment.

(19) Outlining and rehearsing the action to be taken in the event of attack.

78. Shelter Discipline

a. When a shelter is finished, the first man entering it will arrange all equipment in the proper place. The stove, water can, firewood, tools, and rations are placed in the most convenient place by the door of the tent. In a snow shelter, a special storeroom may be dug for these items.

b. In low temperatures, individual weapons are left outside on weapon racks in order to avoid condensation; however, for security reasons, it is advisable to take at least one weapon inside the shelter. Tools should be kept handy for digging out the shelter door after a storm.

c. Before entering the shelter, hoarfrost and snow must be brushed off clothing and equipment. This keep the clothing dry and the shelter clean.

d. To live comfortably in a shelter is not an easy art. Individuals usually are crowded and must keep their equipment orderly and out of the way of other occupants of the shelter. Unnecessary running in and out of the shelter should be avoided whenever possible.

e. The use of fire and lights in the shelter must be carefully supervised. Security, fuel economy, and the prevention of fire and asphyxiation are essential. When wood is available, it is burned in the stoves in place of gasoline. Lamps must be extinguished before retiring for the night. All lamps and cooking stoves must be filled and lighted outdoors. A stand or bracket should be made for the lamps or candles and they should be placed where they are least likely to be knocked over. Sparks on the tent or lean-to roof must be extinguished at once. Smoking while in the sleeping bag is not permitted.

f. As many tasks as possible should be accomplished before retiring in order to conserve time in the morning. All eating utensils should be cleaned, snow melted, canteens or thermos bottles filled, and all weapons should be checked.

g. Upon breaking the bivouac in the morning all personal equipment should be rolled, warm drinks and breakfast should be consumed, and last-minute details accomplished prior to resuming the march.

79. Heat Discipline and Fire Prevention

Heat discipline presents a paramount problem during periods of extreme cold.

a. Overheating the shelter is very common and can and should be avoided. It causes sweating of indivduals and increases the fire hazard.

b. There are many ways to save fuel. Cooking and heating may be combined. The melting of snow and ice uses large amounts of fuel and should be avoided when water from other sources is available. In cooking, liquid fuel is used sparingly. Wood should be burned when available. In extreme cold it may be necessary to keep the fire burning throughout the night in order to keep the men warm, especially when living in temporary shelters which provide little heat. The drying of wet clothing and the providing of hot drinks for combat reliefs are also necessary throughout the night.

c. Fire prevention during both summer and winter seasons is extremely important. The combination of low humidity and the drying effect of continuously heated shelters is conducive to fire. Shifts in wind and the accumulation of frost or soot in the stovepipe lead to backfiring of flaming fuel into the shelter. The excessive spilling of fuel containers, lamps, and candles create additional hazards. The stamping of feet to shake off snow or frost may cause stoves and small heating units to spill and spread fire. The strict enforcement of all regulations is necessary in order to avoid fire hazards. No set rules can be given for each occasion. Common sense in the handling of all kinds of fires, fuels, and flammable materials is essential; alert, wide awake fire guards must be on duty in each shelter at all times when men are sleeping and a fire is burning.

d. A base made from green logs must be placed under the stove if the snow has not been shoveled away from the tent site. Fire reflectors may be used not only to get more warmth, but also to keep the fire burning evenly and to help avoid sparks.

e. Care must be exercised when lighting the gasoline-type stove; it may flare up and either damage the tent or set it on fire. All stove pipes must be cleaned frequently. When using wood as fuel, cleaning must be done every day in order to maintain a good draft and avoid fires in the stovepipes. Detailed instructions for operating stoves are covered in TM 10-735 (Yukon Stove) and TM 10-725 (Stove M1941). Precautions against forest and ground fires in summertime are extremely important. Coniferous forests are highly flammable during the summer season. Ground fires can burn for months in muskeg and are extremely hard to put out. A fire ditch is always dug before lighting a fire. A base of green wood, gravel, or rocks must be used under the fire; the fire must be made on high ground when the forest is dry. Before leaving the camp site, individuals must always be sure that the fire is completely out.

80. Drying Clothes

a. Keeping dry is important in low temperature. At times it is impossible to avoid sweating. The drying of clothes and footgear is therefore a necessity. Every opportunity must be used by each individual to dry his clothing.

b. When drying outside using an open fire, clothes should not be placed downwind from the fire, due to the sparks and smoke. Clothes hung for drying should be frequently checked and not left unattended. Clothing should never be placed too close to the fire or stove in the shelter. Leather items are very vulnerable to extreme heat. Clothing being dried in the shelter is placed on sock lines.

c. The use of a "Christmas Tree" (fig. 36) for drying in the shelter is handy when operating in a wooded area. Branches are cut off a dry or green tree which is then made to stand up in the shelter next to the center pole so that it is in the air current. This offers an excellent place for drying heavy items such as boots and parkas. The Tent, 10-Man, Arctic, is also equipped with strong hooks at the inside peak for suspending lighter weight clothing for drying.

d. Where provided, blanket safety pins can be used for holding drying clothing on a line.

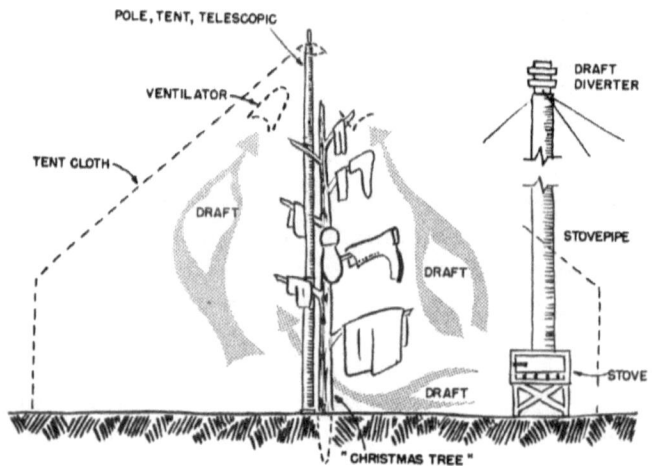

Figure 36. "Christmas Tree" for drying clothing.

81. Sleeping Arrangements in Bivouac

a. When arranging the sleeping procedures in a tent or improvised shelter, the position of every man, especially the position of reliefs for sentries, is planned. Each man must know where his relief is sleeping. Therefore, the floor space is occupied by the individuals in accordance with the duty roster. The number one man sleeps next to the door, number two man towards the rear. In this manner, starting from the door, the relief is easily located without waking up all occupants. The systematic sleeping arrangement will also permit exit from the tent in an organized manner in case of alert.

b. Ground insulation is most important. Often the occupants may have to improvise insulation using all available material. Packboards, snowshoes, man-hauled sleds, and empty cartons may be used. In timbered areas evergreen boughs are especially suitable. On the tundra, dry lichen, grass, or shrubs provide effective insulating material. To make a bough bed, one single bed is constructed for all; the size varies with the number of persons. For improvised shelters, logs approximately three inches in diameter are pegged or fitted around the bough or grass bed. This helps to

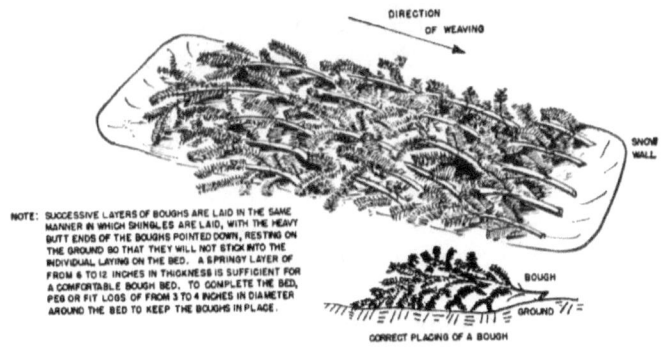

Figure 37. Building bough bed.

keep the boughs in place. If material and time permit, a 6- to 12-inch thick shingled bed made from spruce, fir, or balsam boughs (fig. 37) gives excellent insulation and provides a soft mattress.

c. The tactical situation dictates whether or not sleeping bags are used. The amount of clothing to be worn when sleeping on a bough bed or in the sleeping bag can be best judged by experience and will depend on temperature and the tactical situation. As a minimum, outer clothing is usually removed when the sleeping bag is used. The removed clothing is placed beneath the individual for additional insulation and instant availability. In an emergency it may be necessary to dress in the dark. In the morning all ice and frost is removed and the bag ventilated before rolling it up. Time permitting, it is hung up by the strings and thoroughly dried.

d. When sleeping in a heated tent without a sleeping bag, shoes are usually removed, situation permitting. The outer parka is used like a blanket. The rucksack makes a good pillow. The clothing is always loosened.

82. Water Points and Snow Area Locations

a. Summer. The best available water point is marked. It may be a lake, stream, creek, or spring. Upon arrival at the bivouac, the area is segregated so that the water is used only for drinking purposes. In the summertime all bathing must be done downstream from this point. The garbage disposal must be so located that no trash is thrown into or near the water point; also, the latrines must be located so that the surface water cannot carry impurities to the water point.

b. Winter. During the winter it may be necessary to obtain water by melting snow or ice. When such a source is utilized for drinking purposes, an area should be set aside and restricted to this purpose only. A preferable site is one upwind from the bivouac and isolated from the excreta and garbage disposal areas. If such an area is not available, then snow should be gathered from the branches of trees or lightly skimmed from a carefully isolated area adjacent to the individual shelters. Water obtained in this manner must be boiled for one minute or chlorinated. Chemical sterilization of water under freezing conditions requires a longer period because the disinfecting compounds act with retarded efficiency under such conditions. The time allotted for contact with purification tablets should be two to four times the normal period of one-half hour. Eating ice or snow is unsatisfactory, and may result in painful lesions of the lips, besides the danger of infection.

83. Bough and Firewood Areas

The areas for cutting boughs and firewood should be immediately designated when a bivouac site is selected.

a. Bough Area. The area for cutting boughs for bedding as well as for construction of improvised shelters should be common to all individuals of the group. It is selected in a dense area of woods in which springy, unfrozen boughs are available, and should not be too close to the bivouac site. It is advisable to use sleds in hauling material to the shelter site. Due to the camouflage and track discipline, only one well-concealed trail is used. When cutting boughs, the unnecessary felling of trees should be avoided because trees lying on the ground can be easily observed from the air. Instead of felling trees, only the lower branches should be used.

b. Firewood Area. It is advisable to have the firewood area nearby the area designated for bough cutting so that the same track can be used. Dry, dead pine trees make the best firewood. If no dead trees are available, green birch trees may be chopped; they possess excellent burning qualities even when frozen. The top parts of dead trees should be burned during the daytime, as they give off lighter colored smoke. The lower part of the trunk has more resin and tar, and burns better, but makes more and much darker smoke.

84. Storage

Storage problems in winter are increased by snow, low temperatures, thaws, limited storage space, and the increased problems of transportation. Space in any shelter is limited. Only items which are affected by cold, or which must be immediately available, should be stored inside. All other stores must be concentrated, well marked, covered, and left outside. On the other hand, some perishables which are difficult to preserve in summer may

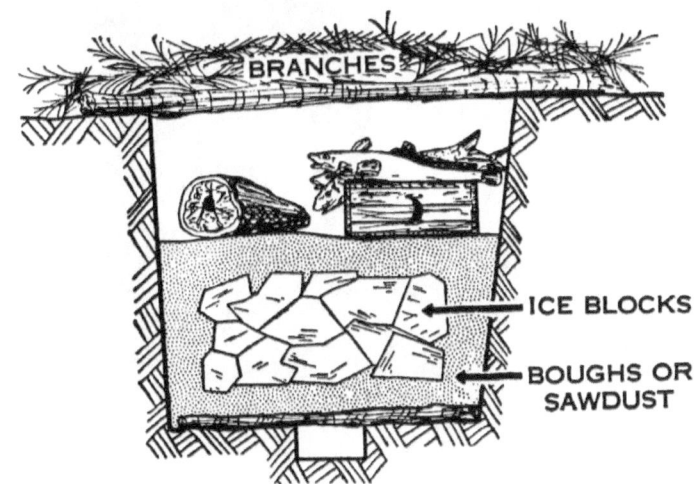

Figure 38. Improvised "ice box."

be kept during the winter months in a natural "deepfreeze" over an extended period of time. In areas where permafrost exists, a hole can be dug or blasted out and then covered with insulating material, such as boughs. A constant low temperature can thus be maintained. In areas where there is no permafrost, an improvised ice box can be constructed as illustrated in figure 38.

a. Rifle Stand and Hanging of Weapons. In wooded terrain a weapon rack may be built from poles placed in a horizontal position and covered with boughs (fig. 39). When boughs are not available, various other materials such as empty cardboard boxes, tent or sled covers, waterproof bags or ponchos will be utilized to protect the weapons from rain, dust, and falling or drifting snow. When weapons are hung outside on stacked skis, or suspended above the snow in some other manner, they are hung with the muzzle down to keep falling or blowing snow out of the barrel and working parts.

b. Ski Racks and Stacking of Skis. Care of skis in the field is highly important because unit and individual mobility depends upon them. If left lying on the snow in the bivouac area, the bindings and running surfaces will freeze and render the skis unusable for a long period of time, or they may be entirely lost under drifting snow. Therefore, the skis and ski poles are placed on an improvised ski rack made of one or two long poles which have been secured between two growing trees in horizontal position (fig. 39). In open areas, skis are simply stuck upright or stacked in the snow as described in appendix IV.

c. Sleds. Sleds are placed on their sides or on end outside. If loaded sleds are left on the snow, sticks, poles, or branches are

Figure 39. Rifle and ski stand.

laid under the runners to prevent them from freezing to the snow. Heavy cargo sleds, 1-ton or larger, must be placed on top of heavy poles or logs due to the fact that sled runners remain hot after extensive usage and tend to settle into the snow and become frozen, making movement of the sled difficult the following day.

d. Vehicles. Oversnow vehicles are driven under a big tree or in lee of a shelter or snow drift. Vehicles should be parked so the least amount of snow can get into the engines.

e. Ammunition and Fuel. Ammunition and fuel are stored separately outside. Ammunition boxes should be stacked off the ground in a dry place and covered with canvas or boughs. In order to locate stacks if snow-covered, a pole should be erected near them. Boughs or poles are placed under fuel containers to prevent them from freezing to the snow.

85. Insect Control

Individuals and small units should have some means of protecting themselves from the swarms of insects during summer operations.

a. Smudges.

(1) Smudges, when properly made, will sometimes furnish relief (fig. 40). For a good smudge, a brisk blaze of dead wood is built up and left burning until a bed of coals is formed. Meanwhile, a supply of additional fuel as well as green ferns, leaves, twigs, and damp and rotten wood is gathered and slowly added to the fire. The dense smoke that results will assist in giving protection from insects. Smudges should not be used in the vicinity of the enemy.

Figure 40. Smudges against insects.

    (2) A bucket smudge built in a pail or pot is some protection against insects. It can be moved easily in case the wind changes or can be put inside a shelter until the insects are driven out.

  b. Screens. The Tent, 10-Man, Arctic, has screen doors and ventilators. The Tent, Hexagonal, Lightweight, does not have a screen door but does have ventilator screens. When a screen is missing, or an improvised shelter is being used, a standardized field-type insect bar may be utilized as described in FM 21–15. An improvised anti-insect screen for a tent may also be made of cheesecloth or bobbinet. Cheesecloth with a circular or oval opening that can be closed with a drawstring may be sewn into the front of the tent. A bobbinet curtain, weighted at the bottom so that it will hang to the ground, serves the same purpose. To be flyproof, the tent must have a wide cloth extension which may be held down with sod. Before entering, the insects are vigorously brushed from the front of the tent and off of the clothing.

  c. Sleeping Bags. If sleeping bags are used, protection may be afforded by covering the face openings with the individual mosquito nets.

  d. Individual Protection.

    (1) Personnel should be thoroughly trained in individual protective measures against insect bites. This training should include instructions in the proper use of insect repellents applied to the skin, aerosols, and proper wearing of the uniform. Insect repellents applied to the skin will give protection against bites from mosquitoes and other biting flies known to occur in the northern areas.

One application of the insect repellent which is presently an item of issue will give protection for as long as five hours under favorable conditions, if it is applied in the proper manner. In applying insect repellent to the skin, all exposed skin surfaces should be covered. Care should be taken that the repellent does not get into the eyes, as it will cause a burning sensation.

(2) If troops are engaged in heavy work and are perspiring, the repellent may have to be applied more often, perhaps as often as every two hours. Aerosols should be used inside tents to kill mosquitoes and other biting insects. Some relief may be obtained by use of aerosols outdoors throughout bivouac areas. Such relief will be only temporary however, and insect repellent will be required for any lasting protection. If clothing is properly worn, i.e., sleeves rolled down and all exposed portions of the skin protected with repellent, the annoyance by mosquitoes will be greatly diminished.

86. Field Sanitation

a. Waste Disposal. Field sanitation in the colder regions is based on the same principles as in temperate climates. The extremes in climate and weather, however, make the problem more acute. The wastes that present constant and real problems are human excreta, garbage, and trash.

(1) In bivouac areas, pit or "cross-tree" type latrines are used for the disposal of human waste (fig. 41). One latrine will usually serve the needs of individuals occupying 3 to 4 shelters, or a unit of platoon size. The latrine is placed downwind from the bivouac, but not so far from the shelters as to encourage individuals to break sanitary discipline. A urinal, designated for each shelter, should be located within 10 to 15 yards of the shelter. A windbreak of boughs, tarpaulins, ponchos, or snow wall should be constructed to protect the latrine from the wind.

(2) When breaking bivouac, the human waste that has accumulated in the latrine is covered with two feet of compacted earth. If, in winter, the latrine has been dug in snow, two feet of compacted snow will be used. All closed latrine sites, tactical situation permitting, will be clearly marked.

(1) During the summer in most of the colder regions a sani-
b. Trash and Garbage Disposal.

Figure 41. Field latrine in the forest.

tary fill can be operated for trash and garbage disposal. A trench is dug to make the fill, which is then covered with earth to prevent attracting insects and rodents.

(2) In winter the edible portion of food waste may be collected in receptacles and disposed of by burial in the snow at a safe distance from the bivouac. Every effort should be made to burn the bulk of the trash and garbage.

(3) All trash and garbage dumps should be marked with appropriate signs to warn troops who might occupy these disposal sites at a later time.

(4) Strict camouflage of all trash and garbage is essential. Dark trash on the white snow is easily seen from the air. Glittering tin cans or bottles may be seen by the enemy. Trash and garbage should be placed under any available cover and camouflaged with snow, branches, or other materials.

c. Rats and Mice. Rats and mice will be found in most of the habitable cold regions of the earth. They are a definite menace to health and property and should be kept under strict control. Rat poisons or traps should be used when available.

CHAPTER 4
SKIING AND SNOWSHOEING

Section I. INTRODUCTION

87. Purpose and Scope

a. The purpose of this chapter is to provide information concerning—
   (1) Techniques used in military skiing and snowshoeing.
   (2) Application of these techniques to increase the oversnow mobility of troops engaged in military operations.

b. This chapter also describes—
   (1) Equipment available for military skiing and snowshoeing.
   (2) Maintenance and care of that equipment.

88. General Considerations

a. The Need for Individual Mobility.
   (1) Warfare in snow-covered areas requires oversnow mobility off the roads. Well-trained ski and snowshoe troops are required on the snow-covered battlefield. Troops on skis are mobile, not roadbound, and are able to move cross-country over all types of snow-covered terrain. They are ideally suited for reconnaissance, security missions, and pursuit. Aggressive action can be carried out against the enemy flanks, rear, or communication lines by lightly equipped, fast-moving troops on skis.
   (2) Deep snow hinders movement on foot. By using snowshoes, individual mobility will be restored to a point approximately equal to that of foot movement on hard ground. By using skis, individual mobility will usually exceed that possible on foot.

b. Need for Certain Techniques.
   (1) During cross-country marches and in combat the soldier on skis or snowshoes encounters various types of terrain. He will move and operate in different weather and snow conditions. Carrying his rucksack and keeping his weapons ready for action, he moves in forests and on open

terrain, uphill and downhill, and often while pulling a sled.

(2) In order to execute his mission without becoming exhausted, the soldier must apply the proper techniques of skiing and snowshoeing required for different conditions under which he will operate.

c. *Use of Oversnow Equipment To Achieve Mobility.*

(1) The means available to the soldier for obtaining oversnow mobility are skis and snowshoes. Using skis, he is normally able to execute long marches more easily and quickly than when using snowshoes. Cross-country movement by soldiers on skis can be facilitated by towing the skiers with vehicles or animals; this cannot be done with soldiers on snowshoes. Snowshoes are more suitable than skis in confined areas, when working close to heavy weapons, or when training time is limited.

(2) Rates of movement over snow-covered terrain cannot be given exactly. They vary in each situation. However, the following rates are given as a rough guide. Rates given are for movement over flat or gently rolling terrain when a man is carrying a rifle and packed rucksack.

Rates of movement	Cross country	Broken trail
On foot—less than 12-inch snow	1–2 mph	1¼–2 mph
On foot—over 12-inch snow	¼–¾ mph	1¼–2 mph
Snowshoeing	1–2 mph	2–2½ mph
Skiing	1–3½ mph	3–3½ mph
Skijoring	N/A	5–15 mph (depending on terrain)

Section II. SNOW AND TERRAIN

89. Snow Composition

Snowflakes are formed from water vapor, at or below 32° F., without passing through the liquid water state. Newly fallen snow undergoes many alterations on the ground. As the snow mass on the ground becomes denser, the snowflakes consolidate and the entrapped air is expelled. These changes are effected by the conditions of temperature, humidity, sunlight, and wind which prevail.

  a. Temperature. In general, the lower the temperature, the drier the snow and the less consolidation. As the temperature rises, the snow tends to compact more readily. Temperatures above freezing cause wet snow conditions. Lowered night temperatures may turn wet snow into an icy crust.

b. Sunlight. In the springtime, sunlight may melt the surface of the snow even though the air temperature is below freezing. When this occurs, dry powder snow is found in shaded areas and wet snow in sunlit areas. Movement from sunlit areas into shaded areas is difficult because the wet snow will freeze to skis and snowshoes. After sunset, however, snow usually freezes and the ease of movement improves.

c. Wind. Wind packs snow solidly. Wind-packed snow may become so hard that skiing or even walking on it makes no appreciable impression on its surface. Warm wind followed by freezing temperatures may create an icy, slippery crust on the snow. Under such conditions, skiing and snowshoeing are difficult. Another effect of wind is that of drifting the snow. The higher the wind velocity and the lighter the snow, the greater the tendency to drift. All troop movement is greatly affected by drifting snow and wind, the effect depending on the direction and speed of the latter. In addition, as the wind increases, the effect of extreme cold (windchill effect) on the body may slow down or temporarily stop movement, requiring troops to take shelter. The snowdrifts created by wind usually make the snow surface wavy, slowing down movement, especially in darkness.

90. Snow Characteristics

The characteristics of snow which are of most interest to the soldier are—

a. Carrying Capacity. Generally, when the snow is packed hard, carrying capacity is greater and movement is easier. Although the carrying capacity of ice crust can be excellent, movement may be difficult because of the slippery surface.

b. Sliding Characteristics. All-important to the skier are the sliding characteristics of snow. They vary greatly in different types of snow and materially increase or decrease the speed of the skier, according to the conditions that exist.

c. Holding Capacity. The holding capacity of snow is its ability to act upon ski wax in such a way that backslipping of the skis is prevented without impairing the sliding characteristics. Holding capacity changes greatly with different types of snow, making it necessary to have a variety of ski waxes available.

91. Effects of Snow and Terrain on Individual Movement

a. Skies or snowshoes are usually employed in military operations when the depth of snow is one foot or more. This equipment is needed in deep snow conditions to provide the necessary oversnow mobility of the individual and the maneuverability of troops.

b. Snow cover, together with the freezing of waterways and swampy areas, changes the terrain noticeably. Generally, the snow covers minor irregularities of the ground. Many obstacles such as rocks, ditches, and fences are eliminated or reduced. Lakes, streams, and muskeg, impassable during the summer, often afford the best routes of travel in the winter when they are frozen and snow-covered. During breakup periods this advantage is reduced, since the snow becomes slushy and the carrying capacity is poor. Even so, skiing or snowshoeing, although slow, is often the only practical way to move during this period. The drop in temperature each night will still freeze the snow surface, creating a good route for a skier or snowshoer during the night and early morning.

c. The effects of snow and terrain on individual movement vary in different areas.

(1) In many areas the terrain is flat or gently rolling. Only a few scattered rock outcroppings, bare ledges, river banks, and shrubs create obstacles. The snow cover in most northern areas is shallow but firmly packed and generally will support a man on foot. However, if soft snow is found, mobility will be increased by the use of skis or snowshoes.

(2) In other areas, terrain varies from vast plateaus to high mountain ranges. Typical features are vast spruce and fir forests, brush areas, tundra, grass and swamps, and numerous lakes and waterways. Skiing and snowshoeing are relatively easy on frozen, snow-covered waterways and swamps. However, the soldier must be accustomed to skiing in woods to obtain concealment and to achieve surprise. In woods the snow is softer, with the result that its carrying capacity is poorer than in open areas, and greater skill is needed to avoid trees and other obstacles. These disadvantages can be reduced by careful selection of routes and by proper trailbreaking. In the spring the woods protect the snow from melting. Therefore, skiing is still possible in woods although open fields may be bare. In autumn the situation is the opposite; the deeper snow is generally found in the open fields, allowing skiing outside the woods earlier than it is possible in the woods.

(3) Mountains present special problems. Their steep slopes demand additional skills from a skier and make move-

ment on snowshoes very difficult. Slopes which are easy to negotiate in the summer often become more difficult and dangerous to cross during the winter because of the deep, usually hard packed and shifting snow cover as well as the added hazard of avalanches. Deep drifts and snow cornices present other obstacles and dangers. On glaciers, snow may make the crossing of crevasses possible, but dangerous (FM 31-72).

Section III. MILITARY SKIING

92. Advantages and Disadvantages

a. Advantages.
  (1) In snow-covered terrain the weakest and the most vulnerable points of the enemy are usually the open flanks, rear areas, and the lines of communication. Attacking, defending, or delaying troops require high degrees of oversnow, cross-country mobility to reach these objectives. Units on skis are the most suitable troops to be used for surprise attack on distant objectives.
  (2) A trained individual or a unit on skis can execute cross-country marches on roadless, variable, and snow-covered terrain more efficiently and quickly than on snowshoes or on foot.
  (3) Skiing over snow-covered terrain by properly trained troops is comparatively less tiring than marching on snowshoes or on foot. Sliding characteristics obtained by the skier increase speed, mobility, and rate of march.
  (4) Due to increased bearing surface, a skier or a unit on skis is able to cross frozen lakes and rivers when the ice will not support a man on foot.
  (5) The use of oversnow vehicles and other suitable means of towing troops further increases their mobility.

b. Disadvantages.
  (1) Individuals require a considerable amount of training before becoming proficient in the use of skis for military purposes.
  (2) Certain terrain features, such as very dense brush and windfall areas, materially decrease the rate of march of a ski unit.

93. Training Objectives

a. General Considerations. A soldier on skis must be capable of moving under control across diversified, snow-covered terrain while carrying the arms and equipment necessary for tactical operations. Since skis are often the most efficient means of transportation in winter warfare, the soldier should be so skilled in their use that skiing becomes a natural method of movement. Since the skiing soldier will utilize his skis for the greater portion of movement over snow-covered terrain, it is important that he acquire good skiing technique in order to be able to move anywhere required both quickly and with the least expenditure of energy. The soldier must develop these techniques so that his movement either uphill or downhill will not become an obtsacle to the movement of his unit. When operating in mountainous areas, the soldier must possess efficiency in both basic and advanced military ski techniques in order to move easily and safely over steep and rough terrain; the soldier must possess endurance and must be in top physical condition.

b. Training Time Required. To walk on snowshoes, one day of instruction is generally sufficient. In a period of two weeks a soldier can be taught enough ski techniques to enable him as an individual to negotiate flat or rolling terrain with greater speed than if he were on foot or snowshoes, but he will not yet be able to operate effectively as a combat skier within a unit. Eight weeks of intensive training are needed in order to become a military skier capable of operating proficiently in any type of terrain. It should be noted that the level of skiing skill developed by the soldier during any period of ski instruction is improved by participating in unit training which is done on skis.

94. Ski Equipment

a. Types of Skis.

(1) The cross-country ski (fig. 42) is usually of laminated wood construction with hickory top and running surface. It is lighter and more flexible than the mountain type and does not have steel edges. The bindings (fig. 43) are suited for use with normal footgear and designed to be easily and quickly put on and taken off. The bindings permit free vertical movement of the heel and assist in normal foot movement. This type of skis is issued for use on level or rolling terrain.

(2) The mountain ski (fig. 44) is wider, less flexible, and heavier than the cross-country type. It is also of laminated wood construction and has steel edges. These metal edges allow the skier to obtain better gripping action in turns, resulting in better control of skis when operating in mountainous terrain.

(3) All skis are painted white and have a hole in the tip through which a cord can be threaded when it is necessary to pull them as ski bundles or as an improvised sled.

Figure 42. Cross-country skis.

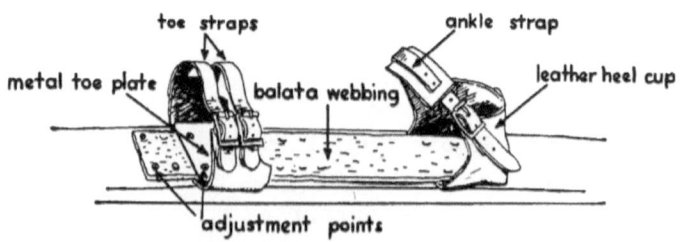

Figure 43. Cross-country bindings.

The skis must be of sufficient length to support the weight of the soldier in loose snow, but not too long or they will be difficult to control. A rough estimate as to the length required can be obtained by the individual by extending his arm straight above his head. The distance between the palm of his hand and the floor will be the approximate length of the ski needed. Cross-country skis are available in two lengths—7 feet and 7 feet 3 inches. Mountain skis are available in four lengths—6 feet 9 inches, 7 feet, 7 feet 3 inches, and 7 feet 6 inches. It should be noted that when these different length skis are depleted in the military system, they will be replaced by skis in seven-foot lengths.

b. *Types of Ski Bindings.*
   (1) *Cross-country ski binding* (fig. 43). This binding is for use when movement is to be made on gentle, rolling terrain and is designed to accommodate all types of cold weather footgear. The binding consists of a strip of "Balata" webbing and a leather toe piece with two straps. The strip of webbing, as well as the toe strap, is secured by a metal plate and four screws to the top of the ski. The webbing has a series of holes in its front portion for adjustment to footgear size. The material used is flexible and allows the heel of the foot to rise freely. The leather heel cup is riveted to the "Balata" strip. The foot is secured to the strip by the ankle strap. The toe of the foot is held in place by the two leather straps of the toe piece.
   (2) *Mountain binding* (fig. 45). The standard mountain binding consists of toe irons, which can be adjusted to fit any size ski boot, and a cable that passes round the boot to prevent the boot from slipping out of the toe irons. The cable is fastened to the ski in front of the toe irons with an added adjustment and clamp which allows the cable to be adjusted for different boot lengths. On the sides of the ski are fixed two small attachments. When it is desired to hold the heel firmly to the ski the cable is passed through these attachments. This position is commonly referred to as the "downhill hitch." When freedom in vertical movement of the heel is needed, the cables are taken out of this downhill attachment and placed in the attachment provided on each side of the toe iron. This position is commonly called the "cross-country hitch" and allows the heels to be raised.

c. *Ski Poles.* Adjustable or nonadjustable tubular steel poles 53 inches long will be furnished for military use. The nonadjustable types will be issued in different lengths. Figure 46 illustrates the different parts of the ski pole. In an emergency, the poles can also be used for many other purposes such as tent poles, markers, or emergency litters.

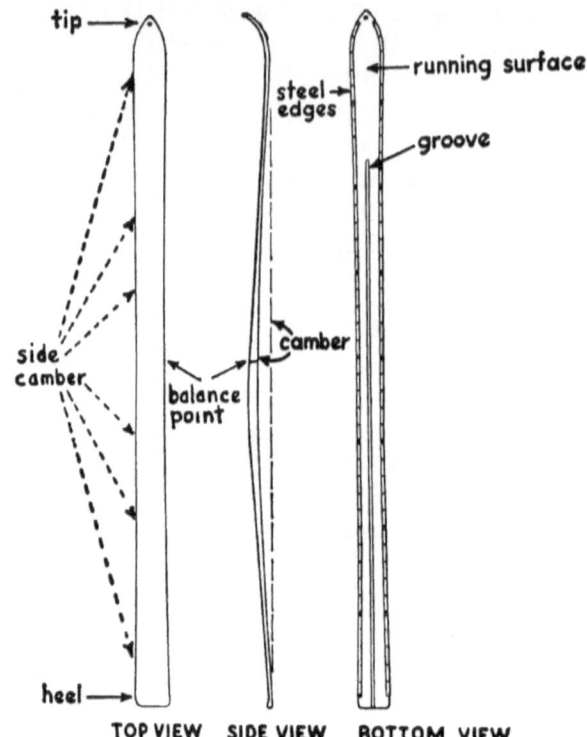

Figure 44. Mountain ski.

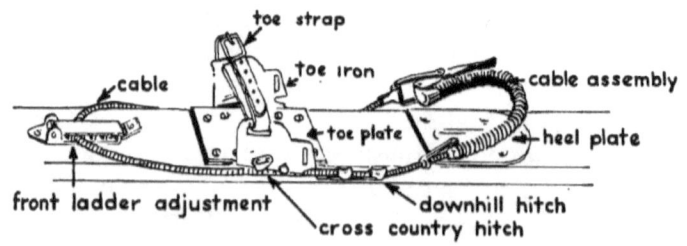

Figure 45. Mountain binding.

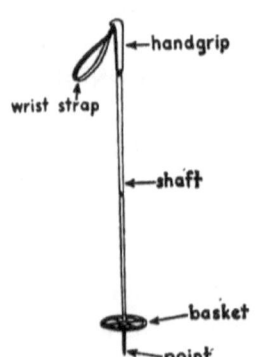

Figure 46. Ski pole.

d. Ski Repair Kit and Emergency Ski Tip. This kit contains pliers, screwdrivers, screws, wire, drill, strips of steel edging, and leather thongs for use in emergency repair of skis, poles, or bindings while in the field. An emergency ski tip is also available. This can be used to repair or replace broken ski tips and allow the individual to continue the march until replacement skis can be obtained. Ski repair kits and emergency ski tips are usually issued to units and are not intended for individual issue. One ski repair kit per rifle platoon and one emergency ski tip per squad is usually sufficient.

e. Ski Waxes. Ski wax is used to obtain the sliding and climbing characteristics necessary for efficient military skiing. Standard items available are Blue, Red, and Orange ski wax. The colors blue, red, and orange are used to distinguish the types of wax and do not necessarily mean that the wax is of that color. The waxing of skis is covered in paragraph 96.

f. Ski Climbers. Climbers are strips of mohair attached to canvas cloth and are supplied in sizes to fit the various lengths of skis. Climbers are attached to the bottom of the skis by means of straps. When attached, the mohair material lies with the ends pointing towards the heel of the skis. Forward movement of the ski does not disturb the material, thereby allowing the ski to slide. Backward pressure, however, causes the material to become roughened, preventing the skis from backslipping. Climbers are used by troops to make the climbing of steep slopes faster and less tiring, providing the ascent is sufficiently long to justify the time required to put them on and take them off. They may also be used to give more traction while pulling sleds, and for descents where sliding is not desired.

95. Preparation of Skis

a. General. Pine tar or ski lacquer is applied to the running surface of the skis to fill the pores of the wood and to furnish a base so that the skis may be properly waxed. They are also applied to the running surface of the skis to prevent moisture from being absorbed by the wood. For military skiing, pine tar is preferred as a base. If this is not available, ski lacquer is a suitable substitute. They must be used separately since they do not mix together.

b. Application of Pine Tar or Ski Lacquer.
  (1) *Preparation of skis.* The running surface must be clean to prepare the skis for pinetarring or lacquering. If the ski has been used, the old base and wax must be removed. The easiest way to accomplish this is to use a scraper and sandpaper. Caution should be exercised to insure

Figure 47. Attaching ski climbers.

that the running surface of the ski is not damaged. Old wax can also be removed by the use of steel wool or a rag soaked in gasoline. If conditions are such that these materials are not available, heat can be used to remove the wax.

(2) *Tarring procedure.* After the ski has been cleaned, a light coat of pine tar is then applied with a soft brush or a rag. If the pine tar is stiff, it should be heated slightly so it can be evenly distributed. Heat is then applied to the running surface to cause penetration of the pine tar into the pores of the wood. The source of heat used may be a blowtorch (fig. 48), one burner stove, or an open fire (fig. 49). To obtain the best penetration, work progressively on one section at a time rather than heating the whole surface of the ski. Care must be taken to avoid burning the wood by application of too much heat. It may be necessary to repeat this procedure several times to obtain a sufficient coating. Excess pine tar is removed during the heating process by means of a rag. When finished, the running surface of the ski should be dry and not sticky to the touch.

(3) *Lacquering procedure.* After the ski has been cleaned, the surface is allowed to dry thoroughly before applying

Figure 48. Heating pine tar with a blowtorch.

the lacquer. The lacquer is applied with a clean brush, rag, or sponge, starting at the tip and working towards the heel using smooth, even strokes in a continuous motion. None of the lacquered areas should be touched until the lacquer is completely dry. This requires several hours. The application should be made at room temperature for best results. At least two separate coats should be applied, making certain that each one is completely dry before the next one is applied. It is recommended that the surface be lightly sanded with fine sandpaper or steel wool, between coats.

96. Waxing of Skis

a. General. The purpose of ski wax is to provide the ski with necessary climbing and sliding qualities to prevent backslip in

Figure 49. Heating skis over an open fire.

various snow conditions. When snow conditions and temperature change, the type and method of application of ski wax will also differ. Before wax can be properly selected and applied, the individual must learn to recognize the different types of snow conditions. It is also valuable to have some knowledge of how ski wax performs in relation to snow. After snow has fallen on the ground, its crystalline structure is continuously altered by the effects of temperature, wind, and humidity. In very cold weather these changes occur much more slowly than when temperature is near 32° F. Therefore, the most important factor of waxing is the effect that temperature has on the character of the snow and its sliding qualities.

b. *Snow and its Effects on Wax.*
 (1) *The effects of snow crystals.* It is important to understand the relation of wax to the holding and sliding capabilities of the snow.
  (a) When the soldier is skiing on the level, or uphill, his body weight gives maximum pressure to the skis. The soft quality of the wax allows the crystal structure of the snow to penetrate the wax under this pressure and thus keep the ski from backslipping. When the pressure is lifted and the ski allowed to slide forward, the penetrating snow crystals will slide free from the surface of the wax, reducing friction. Continuous forward motion, as in sliding, keeps the crystals from penetrating the wax.
  (b) When the skis slide poorly, the following condition generally exists: the snow crystals have penetrated into the wax but will not slide free. This causes clogging of the snow on the running surface and may eventually cause ice to form. Under these conditions the soldier will find that even vigorous sliding of the ski will not break the snow loose from the wax surface. Little or no forward slide can be gained.
  (c) When the skis slide well, but backslip on the level and when moving uphill, the following condition exists: the snow crystals are not penetrating the wax. The soldier will find he has excellent sliding when going downhill, but climbing uphill or skiing on level ground is very exhausting because of backslip.
 (2) *Classification of snow.* Snow is classified here into four general types. This classification is intended to assist the soldier in snow identification, choice of wax, and its proper application under these different conditions.
  (a) *Wet snow.* This type of snow is mostly found during the spring, but it may also occur in fall or later winter, particularly in regions of moderate climate. This type of snow can be readily made into a heavy, solid snowball. In extreme conditions, wet snow will become slushy and contain a maximum amount of water.
  (b) *Moist snow.* This type of snow is generally associated with early winter, but may also occur in midwinter during a sudden warmup period. This type of snow can be made into a snowball, but will not compress as readily or be as heavy as a wet snowball. It will have a tendency to fall apart.

(c) *Dry snow.* This type of snow is generally associated with winter at its height, but it can occur in late fall as well as in spring, when abnormally low temperatures occur. This snow is light and fluffy. It cannot be compressed into a snowball unless the snow is made moist by holding it in the hand. At extremely low temperatures, such as those found in the Arctic, this snow is like sand, and has very poor sliding qualities.

(d) *New snow.* This is snow which is still falling or has recently fallen on the ground, but has not been subject to changes due to the sun or temperature variation. It can be wet, moist, or dry in nature.

c. *Proper Selection and Application of Waxes.* There are three types of standard issue waxes; Red, Blue, and Orange wax. These waxes are easily identified by their containers, which are marked with the appropriate color and instruction. The proper application of all waxes is important to achieve desired results whether they be sliding action or traction. As a general rule red wax, properly smoothed out and polished, provides the best sliding surface for all types of snow and additionally provides an excellent base for application of other waxes. To provide traction, varying amounts, combinations, and methods of application of other waxes are used. When pulling a sled or carrying a heavy load, the soldier may have to apply heavier coats of blue or orange wax to insure required traction.

(1) *Wet snow.* For wet snow, having a high water content, orange wax is used. This wax is very soft and easily applied. Apply a thin, even layer over the entire running surface of the ski, adding more layers according to existing snow conditions. Generally, the wetter the snow the heavier the coat of wax applied. When wet snow freezes and gets icy, orange wax may also be used. In movement from sunny to shaded areas where the snow is drier, the skis may ice up as a result of freezing of the snow crystals sticking to the wax. If this condition occurs, it may be corrected by adding to and mixing red wax with the orange wax. If backslipping occurs, apply a heavier layer of orange wax for approximately 18 inches along the running surface of the ski directly under the foot.

(2) *Moist snow.* Blue wax is used in a thick application for this type snow. If the snow is very damp, a mixture of both blue and orange wax can be used. In this case, apply the blue wax first; then mix the orange wax in very lightly. If snow balls up under the ski, a light application of red wax will correct this condition. If backslip-

ping occurs in damp snow, apply a heavy layer of blue wax for approximately 18 inches along the running surface directly under the foot. This layer is not smoothed out.

   (3) *Dry snow.* Blue wax, applied in moderate thickness and smoothed out, is used for this type of snow. For new, dry snow, this application should be thinner and smoother than for old dry snow. If backslipping occurs in dry snow, apply a moderate amount of blue wax for approximately 18 inches along the running surface directly under the foot, and smooth out enough to eliminate rough spots. Do not use orange wax, either alone or mixed with blue, to increase sliding of the skis. In lower temperature brackets, a very thin and well-polished application of blue wax is necessary. As even lower temperatures are experienced a very thin and well-polished layer of red wax is necessary. In extremely cold conditions, where sliding qualities of snow are the poorest, it may be found that a base of pine tar without any wax will perform most satisfactorily.

  d. Waxing Procedure.
   (1) Whenever possible, the waxing of skis should be done before the march when shelter and heat are available, as the running surface of the ski should be warm and dry to obtain best results. When on the march, ski wax should be carried in the pockets, if possible, so that body heat will keep the wax soft and easy to use. If the skis need waxing during the march, the running surfaces are dried as much as possible by the use of paper or dry mittens. When temperatures are not severe, rubbing the bare hand vigorously along the surface will help get rid of a lot of moisture. Whenever possible, old wax should be removed before rewaxing skis.

   (2) To apply, coat the entire running surface with wax until it has the desired thickness. After this, smooth the wax down by rubbing it with the palm of the hand (fig. 50). When heat is available, this process can be made easier by warming the wax already applied. Work progressively on a section at a time. It is normally best to work from the ski tip towards the heel, leaving minute ridges that will not impede sliding action but which will provide some additional traction. If the waxing is done in a shelter, or heat is used, the skis should be allowed to cool to outside air temperature before being used. Do not place them with the running surface on the snow immedi-

ately after waxing as the snow may stick and freeze to the running surface. For the same reason protect them against wind-driven snow. To insure that wax is properly chosen and applied, the skis should be tested before being used.

97. Care of Ski Equipment

a. General.

(1) A broken ski or binding may put a soldier at the mercy of the enemy and the elements and prevent him from accomplishing his mission. If the soldier keeps his skis and equipment in good condition, he will find that ski marches are easier and less tiring and that he will not be the cause of any unnecessary delays and halts by his unit.

Figure 50. Smoothing wax with the palm of the hand.

Care of ski equipment is the responsibility of the individual soldier—he must check it before starting out on a mission, during breaks, and when in bivouac. At least once a week the ski equipment should be thoroughly checked by appropriate officers. During combat the inspection must be done whenever the situation permits.

(2) Skis must be checked for proper base of pine tar, evidence of possible warping and splitting, loss of camber, and defective edges. At the same time, bindings must be checked for worn straps, missing rivets and screws, improper adjustment, and broken cables. Ski poles should be checked to insure that wrist straps, hand grips, baskets, and points are firmly fastened and that no breakage has occurred.

b. Daily Care.

(1) After each day's use, the skis and the skiing equipment should be checked and necessary repairs made by the individual as follows:

(*a*) *Skis.* Remove any snow or ice that has frozen to the ski. This may be done with heat. If heat is not available, this can be done with a mitten, wooden stick, or piece of metal. Check the heels and tips of the skis for cracks. Badly cracked skis must be replaced, as they are weakened and break easily. At the same time, check for and replace defective or missing edges and screws. The condition of ski bottoms is then checked and, if needed, additional pine tar or red base wax is applied. The surface waxing for the next day's march is deferred until snow conditions are determined in the morning. After maintenance of skis is completed, place them in a ski rack (fig. 51), which should be available outside the unit living quarters. Under field conditions, skis are placed in an improvised ski stand, leaned against trees, or stacked.

(*b*) *Bindings.* Insure that all straps, buckles, cables, screws, and rivets are present and in good condition. Replace parts which are unserviceable. If necessary, readjust the fit of the bindings.

(*c*) *Poles.* Check wrist straps, hand grips, shafts, baskets, and points to insure that they are in good condition. Broken parts should be replaced at the first opportunity. Temporary repairs can be made with wire, cord, or tape.

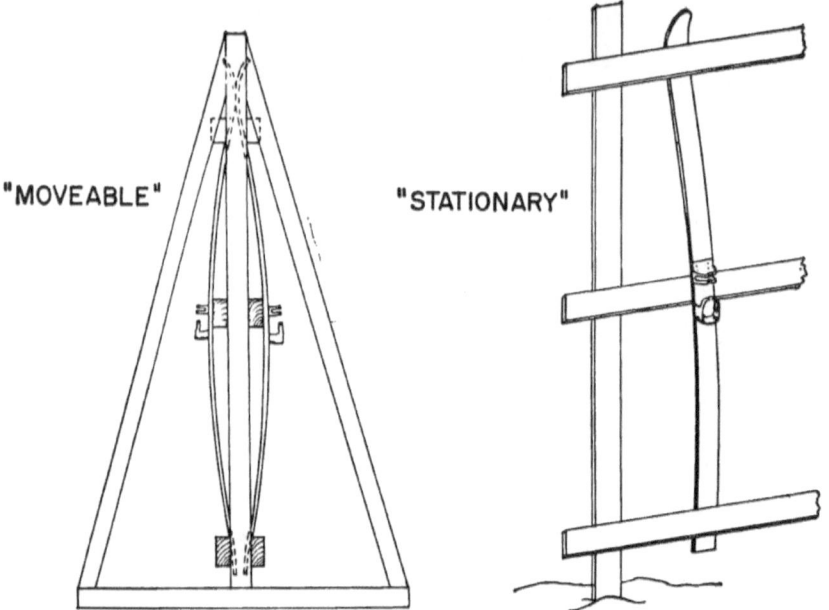

Figure 51. Two types of ski racks.

(2) When snow cover is comparatively thin, be careful not to damage the skis while skiing in rocky or stumpy terrain. Sometimes there is water under the snow cover on frozen rivers or lakes. Try to cross them at a dry place; make an improvised hasty bridge from trees or boughs, if time permits. If the skis become wet during a crossing of water, the ice which forms on the skis must be removed after reaching the bank. A long march or a change in temperature may require waxing of skis during the march. When skis are removed, do not leave them on the snow. It may stick and freeze on the running surface. Remove the snow from the skis and stack them beside the ski tracks; or lean the skis against a tree. A ski stack can be built for each squad.

c. *Repair.*
 (1) *General.* Repair of unserviceable ski equipment requires qualified personnel with necessary tools and facilities. Therefore, the soldier will only be permitted to make emergency repairs such as replacing bindings, cables, screws, and steel edges.
 (2) *Emergency repair.* The repair of ski equipment under field conditions is emergency repair. In many cases broken skis or worn out parts of ski equipment must be replaced. To facilitate this, the following arrangements are necessary:

Figure 52. Attaching an emergency ski tip.

    (a) Every unit should have replacement skis, bindings, and poles. There should also be available ski repair kits, pine tar or lacquer, and waxes.

    (b) Every squad should have one emergency ski tip (fig. 52) and each platoon, one ski repair kit.

    (c) Every man should have the following in his possession at all times:
        1. Emergency thong.
        2. Pocketknife.
        3. Piece of light wire (malleable).

  (3) *Combat repair.* During combat, the most suitable time for maintenance and repair of skis and ski equipment is when the unit is in reserve.

d. Storage.

  (1) Proper storing of skis and skiing equipment is most important during off seasons. Improper care will damage this equipment, making it unserviceable.

  (2) When the skiing season is over, skis and poles are turned in by the using unit for storage. Before doing so, the skis must be cleaned and old waxes removed.

  (3) Skis and poles are then checked thoroughly. Those in good condition are separated from those in need of

repair or salvage. Necessary repairs are made. Ski bindings are not removed. All skis should be pinetarred or lacquered. If needed, skis are repainted. Skilled personnel are needed for repairing skis and poles and for preparing them for storage.

(4) In further preparation, the skis are tied together by pairs according to their unit mark. A piece of string or cord is used to tie the skis together at their tips and heels. A wooden block is then placed between the skis at the toe plates. This block should be of such dimension that a 1½-inch spread on this part of the skis is obtained (fig. 53). After being blocked, the skis are stored in either a vertical or horizontal position. If the skis are stored horizontally, they should be supported at both ends and at the middle. The storage room should be dry with an even temperature and good ventilation.

(5) Ski poles are checked, repaired, and reconditioned. Nonadjustable poles are matched according to length and tied with rope so that each bundle contains 10 pairs of poles of equal size. The ski poles can be kept in the same storage room as the skis.

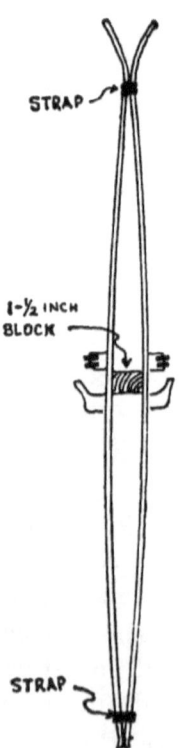

Figure 53. Blocked skis for storage.

98. Basic Movement

a. General. In moving on skis for the first time, most beginners find that skis are awkward to handle due to the difficulty of obtaining the necessary balance and coordination. To overcome these difficulties, the first instructional phase is devoted to walking on level ground in order to obtain the balance, correct body position, coordination, and rhythm necessary in skiing. In addition, this basic movement is a means of forming the foundation for further instruction. Ski drill techniques are covered in appendix IV.

b. The Walking Step.

  (1) *Use.* This is the simplest movement in skiing and is used as the basic step in forward motion. In military skiing, its application is for situations where walking or climbing is necessary. On level ground, sliding action of variable degrees can be obtained.

  (2) *Technique.*

    (*a*) From the position of attention on skis (app. IV, par. 17) left unweighted ski is slid flat over the surface of the snow and straight forward as in normal walking.

    (*b*) At the same time, both knees are bent and the body weight is gradually shifted onto the advanced foot. The heel of the rear foot is raised.

    (*c*) The right ski pole is moved forward and the basket is placed close to the right ski, towards the tip, with its shaft leaning to the front.

    (*d*) A push to the rear with the pole is made, assisting in the forward body motion.

    (*e*) The above motion is repeated with the right ski.

    (*f*) On level ground the skis are kept flat and parallel.

    (*g*) The skis are not lifted off the snow, and the weight of the skis is carried by the snow.

c. Skiing Without Poles. The soldier will find that in performing duties, especially in combat, he will be required to ski either with poles carried in one hand or without poles. For this reason, it is important that he practice all techniques with and without the use of ski poles. This is especially important in the beginning stages of skiing, as practice without ski poles will aid in learning proper transfer of body weight, balance, timing, and control of the skis.

99. Step Turn

a. Use. The step turn is the simplest means of changing direction from a standing position. It is particularly valuable in brushy and wooded terrain (fig. 54).

1 To begin turning, the right ski is placed approximately 45° to the right.

2 Left ski placed parallel with the right to complete the first cycle.

3 If the further turning is desired the right ski is again placed approximately 45° to the right.

4 Left ski placed parallel with the right to complete second cycle.

Figure 54. Step turn.

b. Technique.

(1) From the standing position the right (left) ski tip is raised, the ski is rotated to the right (left) side, using the heel of the ski as a pivot.

(2) The ski is placed on the snow and the body weight shifted onto it.

(3) The left (right) ski is moved along side the right (left) in the same manner.

(4) Each pole is raised, moved, and placed with the corresponding ski (i. e., right ski, right pole).

(5) The same movement is repeated until the desired direction is obtained.

(6) In confined areas it may be necessary to use the tip of the ski instead of the heel as a pivot point. In turning to the right (left) the heel of the left (right) ski is raised off the snow and moved to the left of its original position. Then the right (left) ski is moved alongside the left (right) ski and this sequence repeated until the desired direction is achieved.

100. Kick Turn

a. Use. The kick turn is a method for reversing the direction

1 Preparatory movement for a turn to the left.

2 Left ski kicked up to a vertical position for second movement.

3 Left ski placed parallel to right ski in the new direction.

4 Right ski brought parallel to the left ski to complete the turn.

Figure 55. Kick Turn.

of a skier when in a standing position. It is used on both flat and steep terrain. In combat, it is also useful to conceal a change of direction in a ski track (fig. 55).

  b. Technique.

    (1) Beginning in the standing position with skis level, the left (right) pole is placed alongside the left (right) ski approximately 18 inches to 24 inches in front of the toe of the foot. At the same time the right (left) ski pole is placed alongside the right (left) ski about 18 to 24 inches behind the heel of the foot.

    (2) The right (left) leg is swung forward and upward until the ski is momentarily perpendicular, its heel alongside the tip of the left (right) ski. To obtain sufficient momentum for this movement, a preliminary backward movement of the right (left) ski should first be made.

    (3) The right (left) ski is then pivoted on its heel and lowered, pointing in the opposite direction and parallel to the left (right) ski.

    (4) The body weight is shifted to the right (left) ski, bringing the left (right) ski and pole around and alongside the right (left) ski in the new direction, placing the ski pole in the snow.

    (5) On a gentle slope the procedure is the same, except the uphill ski should be turned first.

    (6) On a steep slope the skis are placed horizontally across the slope; the movement is the same as described above with the exception that both ski poles are placed in the snow above the skis and the downhill ski is turned first.

101. One Step

  a. General. The basic movement of the one step is the walking step. Forward motion and glide are increased when the skier applies more effort to his step. This added effort is obtained by a lunge coordinated with an increased push from the poles.

  b. Use. The one step is the most widely used of all skiing steps. It is applied under all types of snow conditions on level ground and when skiing slightly uphill or against the wind (fig. 56).

  c. Technique.

    (1) The one step is started by a forward lean of the body, with well bent knees and ankles. The feet are kept flat and the body weight is on the right ski, from which the initial movement (lunge) is made.

A step with the right ski has been completed. Right ski is gliding left ski being brought ahead. Left pole placed in snow.

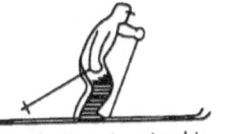
Left ski moving ahead in preparation for lunge from right ski onto left ski.

Lunge beginning from right ski onto left ski.

Lunge progressing onto left ski push being placed on left pole.

Lunge and push with pole fully completed. Left ski gliding. right pole in position to place in snow. Before glide is lost the right ski is brought ahead ready for the next cycle.

NOTE: In this and the illustrations which follow the lined areas denote the proportion of body weight to be placed on the indicated leg or ski.

Figure 56. One step.

(2) The left, unweighted ski is slid flat and straight forward by a springing motion from right ankle, knee, and hip, straightening the body and transferring the weight to the left sliding ski.

(3) Straightening the right knee and pushing off from the ball of the right foot completes the lunge.

(4) The body weight is kept on the sliding (left) ski. The right leg is relaxed and moves the ski forward in preparation for the next step. As this leg reaches a position approximately alongside the left leg, the next step is made with the right ski in lunging from the left leg.

(5) When using the poles, the lunge is executed as above except that as the left foot is slid forward the right ski pole is swung straight to the front and placed towards the tip of the right ski or, when the right ski is slid forward, the left ski pole is brought forward.

(6) The slide is increased by a push with the ski pole. The ski pole is leaned slightly to the front and the elbow kept close to the body.

(7) The pushing action of the ski pole is increased progressively by the muscles of arms and shoulders. The push is finished off by a sharp straightening of the arm for added power. When the push has been completed the arm is relaxed and brought forward close to the body in preparation for the next poling action.

(8) During the coordinated movement of poles and lunge, correct timing and a long glide are emphasized. All motions are rhythmic and fluent. Poles are used in a relaxed manner and the pressure of pushing is allowed to come on the wrist strap.

102. Two Step and Three Step

a. Use. This step is used to attain a longer and faster glide on the level and on gentle slopes. It is also used as an aid through dips and over bumps.

b. Technique. The technique of the two step is a combination of an accelerated walking step and a one step. In the two step the push is obtained by the use of double poling (fig. 57).

(1) From a standing position with the knees slightly bent, a walking step is made with the left ski to start the body in motion initially.

(2) A lunge is then made from the left leg, in a continuous rhythmic motion, to produce a long glide on the right ski.

(3) The two steps are now completed. While gliding on the right ski, the left ski is brought slowly forward and even with the other ski in preparation for starting the next two step. This action should be started before the momentum of the glide has been lost.

(4) As the first step is made, both ski poles are brought straight to the front in a comfortable reach and set into

Figure 57. Two step.

the snow alongside the skis in coordination with the lunge of the second step.

(5) The pushing action with the poles is applied in the same manner as described above in using one pole. As the poles leave the snow, they are brought forward in a straight line in preparation for the execution of the next step. It is most important to time this motion properly to coordinate with the next lunge.

c. Three Step. In addition to the two step, a three step may be used when sliding is good. It has the advantage over the two step of allowing lunging from alternate feet. This step is made in the same manner as the two step except that two initial walking steps are taken before the lunge.

103. Variations and Applications of Ski Steps

a. In long, cross-country movement, particularly when skiing with pack and rifle, it is most important to apply techniques properly according to the terrain to insure that energy is spent wisely and conserved as much as possible. To this end, the individual must attempt to obtain as much glide as possible from his skis during each step. Although lasting only for a short moment, the glide will allow the skier to rest temporarily. In addition, all movements must be made in a relaxed manner, which necessitates continuous individual training. The constant use of the same step is monotonous and increases fatigue. To avoid this, various steps are used temporarily. The same effect is also necessary in poling. In order to rest arm and shoulder muscles, a series of steps may be made without poling. In the one step, for instance, the first two steps can be made with the aid of the poles and the following two steps then executed without using the poles. Any additional combination of steps and poling may be made at one's discretion for the same reason, placing more emphasis on leg than arm work, or vice-versa.

b. In bumpy terrain, ski steps and poling may be used individually or in various combinations to provide a strong push-off to provide the skier with sufficient glide for a continuous motion through a dip and over a bump. When a series of bumps and dips is encountered, the poling action is generally applied on the crest of the first bump in order to obtain sufficient momentum to reach the top of the next bump in a continuous glide. A step supported by double poling may be applied when skiing through the dip. There are other situations where double poling may be applied simply to gain or increase forward motion of the skis without taking a step.

104. Falling

a. General. In military skiing there are two types of falls, controlled and unintentional.

  (1) *Controlled falls.* The controlled fall has definite value. It can be used to avoid excessive speed or to avoid hitting obstacles. The controlled fall can be done safely only at slow to moderate speeds. It is used to take cover quickly, assume a firing position or for a quick stop to avoid hitting an object. When properly used, it can be accomplished without injury to the individual.

  (2) *Unintentional falls.* Unintentional falls are undesirable and may cause serious injury. Other undesirable results of an unintentional fall are increased fatigue, possible frostbite, and holes in the snow which may cause other skiers to fall. Factors which may contribute to unintentional falls are poor skiing ability, lack of control, snow conditions, and excessive speeds.

b. Technique of Falling.

  (1) If a fall is imminent, an attempt is made to relax, lower the body, and to land sideways and to the rear.

  (2) The body is stretched to full length, arms and legs are extended, ski poles pointed to the rear and skis kept close together (fig. 58).

  (3) The impact of the fall should be absorbed by the hips or buttocks.

  (4) The unintentional fall is avoided as much as possible. It is often avoided by the correction of a faulty ski or body position.

  (5) Landing directly on a knee or hand must be avoided since the resulting blow may cause serious injury.

105. Recovery

a. To recover from a fall, the skier must first figure out what to do before attempting to rise. A little planning will save time and energy.

b. If necessary, the pack and other restrictive loads are removed.

c. Skis are untangled and brought parallel, feet together. Knees are pulled up to bring the skis close to the body. The body is then moved forward and raised, pushing with the pole if assistance is needed.

d. To use the ski poles, both hands are first removed from the straps. The poles are then placed together with baskets in the

1   Falling on the slope.

2   Recovery from a fall on the slope.

Figure 58. Falling and recovery.

snow slightly to the rear, grasped with one hand above the basket, palm facing downward, and with the other hand close to the top, palm facing upward.

e. The procedure for recovery from a fall on a slope is the same except that the skis are placed below the body and horizontal to the fall line. To obtain this position it may be necessary to roll onto the back, lifting the skis in the air and then placing them in the proper position. Poles are then used as described on the uphill side (fig. 58).

106. Straight Uphill Climbing

a. Use. Straight uphill climbing is a method of ascending gentle and moderate slopes.

b. Technique.
  (1) Take the first step as in walking, the body leaning forward with knees well bent.
  (2) On gentle slopes, slide the skis forward without lifting them from the snow. On steep slopes, more knee bend is needed and a complete transfer of body weight is made. It may become necessary to lift the ski as the step is made and place it with a pressing action against the snow. This will give the ski better holding qualities.
  (3) Use the ski poles to assist the body in its uphill movement and to minimize backslip.

107. Sidestep

a. Use. The sidestep is an effective method of climbing a short, steep slope. Where space is confined, it may be the only practical means for ascending slopes. It is also useful for stepping sideways over logs, stumps, and other obstacles.

b. Technique.
  (1) The skis are placed together and horizontally across the slope (fall line). To prevent slipping sideways, the uphill edges of both skis are forced into the snow by pushing both knees forward and toward the slope. Avoid leaning into the slope. Initially, the weight of the body is placed on the lower ski.
  (2) The uphill ski is lifted in a sideways step up the slope (fig. 59) and the body weight placed upon it. The upper ski pole is moved at the same time and placed alongside this ski.
  (3) The lower ski is then moved up as close as possible to the uphill ski, while the skier is supported by a push on the

Figure 59. Sidestep.

lower pole. This pole is then brought up and placed alongside the lower ski. This completes one cycle of the sidestep. Merely repeat until the desired elevation is reached.

108. Uphill Traverse

a. Use. This method of climbing is also used when the slope becomes too steep for going straight uphill. Although a traverse generally involves a zigzag route, it will often be the least tiring method of ascending, thereby conserving time and energy.

b. Technique.
  (1) An angle of ascent is selected which will allow climbing without backslip.
  (2) The skis are edged on each step and the ski poles are used as in straight uphill climbing.
  (3) In changing the direction of ascent a kick turn or a herringbone turn, (par. 110*d*) can be utilized. Long traverses should be used whenever possible, since elevation is gained more effectively and with less expenditure of energy in this manner.

109. Sidestep Traverse

a. Use. This step is a combination of a sidestep and the uphill traverse. It allows greater vertical climb in each traverse.

b. Technique.
   (1) The movement is the same as in the uphill traverse, except the ski is raised slightly and placed uphill as it is brought forward with each step.
   (2) The skis are kept parallel and edged, as in the sidestep.
   (3) The ski poles are moved in the same sequence as in the sidestep.

110. Herringbone

a. Use. The herringbone is used to climb short, moderate, or steep slopes. It is quicker than the sidestep. It is more tiring and should be used only for relatively short ascents.

   b. Technique.
   (1) The body is faced uphill with skis spread to form a wide V. This is obtained by spreading both ski tips outward. The skis are edged sharply inward, to prevent backslip, by bending the knees forward and inward (fig. 60).
   (2) The first step is made by placing the weight on one ski, raising the other slightly above the snow and moving it forward and upward. This ski is then placed in the snow, edged inward, and the body weight transferred to it. The other ski is then moved in the same manner and placed slightly ahead.
   (3) The ski poles are used in the same sequence as in a

1 Herringbone.

2 Half herringbone

Figure 60. Herringbone and half-herringbone methods of ascending.

walking step, except they are placed to the rear of the body and to the outside of each ski to act as a brace and to aid in the climb.

c. *Half-Herringbone.*
   (1) *Use.* The half-herringbone is a variation of the herringbone technique and is used to aid in preventing backslip on gentle to moderate slopes in both straight uphill climbing and traversing (fig. 60).
   (2) *Technique.* The half-herringbone is executed with one ski in the herringbone position, the other pointing in the direction of movement. The poles are used for support to prevent the ski pointed uphill from backslipping while the other ski is advanced. The amount that this ski is angled and edged into the slope is increased as the steepness of the slope increases.

d. *Herringbone Turn.*
   (1) *Use.* The herringbone turn is a method of changing direction while traversing a slope, while climbing, or when in confined areas where a kick turn may be difficult to use. It is also used to change direction from a herringbone position.
   (2) *Technique.* From a traversing position the upper ski is moved first in the desired direction, using its heel as a pivot point. This ski is then placed into the snow, as in a herringbone step, with the full body weight on it. The other ski is moved up in the same way and placed into the snow. This brings the skier into a herringbone position. Both poles are held to the rear to brace the body during this movement. This cycle is repeated until the lower ski has reached the desired direction. The upper ski is brought parallel with the lower ski into a traversing position again, completing the herringbone turn.

111. Straight Downhill Running

a. *Use.* Straight downhill running is the first technique learned in skiing downhill. It provides the individual with the balance which he must have before he can effectively descend a slope or learn more advanced techniques for a descent. Although it is the fastest means of descending, speed must be kept within the capabilities of the skier (fig. 61).

b. *Technique.*
   (1) In a normal standing position with skis flat and parallel, one ski is advanced a few inches.

Figure 61. Straight downhill running.

(2) Body weight is evenly distributed on both skis. The knees are bent and pushed forward from the ankles, keeping the heels flat on the skis.

(3) The body is leaned slightly forward in a natural position, head up and without a forward bend of the body at the waist.

(4) Ski poles are held pointing to the rear with baskets above the snow. The arms are bent slightly at the elbows with hands to the front.

(5) Body and arms are kept relaxed. Knees are kept supple to act as shock absorbers. The skier must be alert at all times.

112. Traversing Downhill

a. Use. Traversing downhill is the method most commonly used in a descent, either being used by itself or in combination with other techniques. An individual who has learned the techniques and has chosen a gradual route of descent can, in combination with a kick turn, travel over a great variety of terrain (fig. 62).

b. Technique.

(1) The basic position is that of straight downhill running, except that the uphill ski is always slightly advanced and more of the weight is on the lower ski.

(2) Stand directly over the skis and avoid leaning into the slope. This action edges the skis for normal descent.

(3) If more edging is needed, it is controlled by the knees and is kept even and constant.

(4) The ski poles are held as in the straight downhill positions.

Figure 62. Traversing downhill.

113. Braking Methods

a. Use. The snowplow is a means for controlling and slowing down forward motion in all types of terrain. In gentle or moderate terrain it can be used for stopping. The snowplow uses fundamental positions which are employed for furthering other skiing techniques (fig. 63).

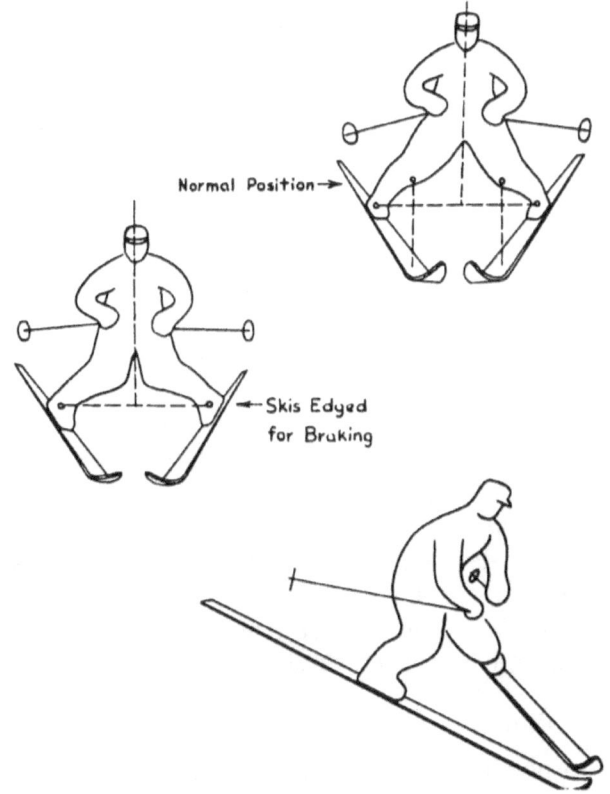

Figure 63. Snowplow positions.

b. Techniques.

   (1) *From straight downhill running position.*

      (*a*) To move into a snowplow, both heels are pushed outward evenly, keeping the ski tips even and close together, forcing the skis to form a wide V.

      (*b*) The body weight is kept even on both skis. The knees are bent well forward in the direction of the ski tips, causing the skis to be edged slightly inward. The heels are kept constantly on the skis.

      (*c*) The upper part of the body and the ski poles are held as in the straight downhill running position.

      (*d*) To increase the braking action, the skis are moved into a wider V and edged more.

   (2) *Half snowplow.*

      (*a*) When only one ski is brought into snowplow position, this is referred to as a half snowplow. The half snowplow is used in confined areas and in traversing where a full snowplow is impractical for braking action. It is also used in conjunction with basic and advanced turns.

      (*b*) This motion is executed by pushing only one ski outward in the snowplow position described above. Braking action is controlled by the degree of weight placed on this ski and the amount of edging applied.

   (3) *The snowplow position while traversing.*

      (*a*) To move into a snowplow from the downhill traverse position, the body weight is shifted momentarily to the uphill ski. The lower ski is then moved downhill into a half snowplow position, keeping the ski tips even and edging slightly. The body weight is then transferred back onto this lower ski. Additional braking action can be obtained by increasing the edging of this ski. To complete the snowplow, the upper ski is flattened and pushed uphill in full V (fig. 64).

      (*b*) If it is desired to continue traversing in a snowplow position, most of the weight is kept on the lower ski. Braking action is increased by a wider spread of the skis and increased edging of both skis.

114. Ski Pole Riding

a. Use. Ski pole riding is a braking method which is sometimes necessary to use in confined areas where ability to control descent

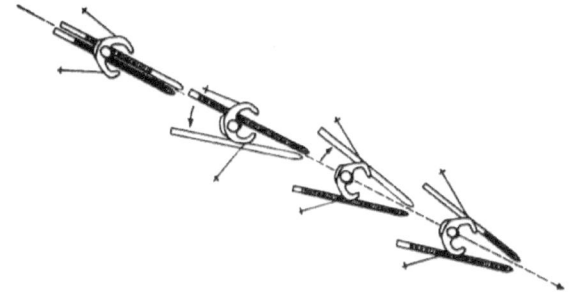

Figure 64. Method of getting into a snowplow position from a traverse.

is limited by snow conditions, terrain features, or obstacles. Two different methods are used (fig. 65).

b. *Technique.*
   (1) Poles are kept together and to the rear and held between the legs for vertical descents.
      (a) From a straight downhill running position the lateral spread of the skis is increased and both hands are removed from wrist straps. Both poles are held together and placed to the rear between the legs and the heels of the skis.
      (b) The body is placed in a squatting position with the weight over the skis and one hand grasping the pole

1 Straight downhill.   2 On a downhill traverse.

Figure 65. Ski pole riding.

handles in front of the body with the palm facing upward, while the other hand is placed to the rear, grasping the shafts above the baskets, palm facing down.
- (c) Control of descent is obtained by applying the required pressure on the ski poles to force the baskets into the snow.
- (d) The braking action may be increased by using the half snowplow or snowplow position.
- (2) Poles together and on either side of the body for traversing.
    - (a) From a downhill traversing position both hands are removed from wrist straps and the poles are held together on the uphill side.
    - (b) The hand on the uphill side grasps both pole shafts near the baskets with palm facing down and the other hand is held near the pole handles, palm facing upward.
    - (c) The uphill arm is braced tightly against the hips to increase the braking action.

115. Sideslipping

a. Use. Sideslipping is a braking method used in descending slopes at all speeds. It is especially useful in confined areas and in steep terrain where the snowplow or pole riding is impractical. It is the least tiring method of braking. In addition, it employs a sliding action which is characteristic in advanced turns.

b. Technique.
- (1) A downhill traverse position is assumed. The edging of both skis is decreased by bending both knees well forward and slightly outward. This minimizes the holding power of both ski edges so that gravity will cause the skier to slide sideways down a hill (fig. 66).
- (2) Care must be taken that the weight is kept well centered on the skis and that the lower ski pole is not placed in the snow during the sliding action. The uphill pole may be used for balance.
- (3) By shifting the body weight in front of the center of the skis while sideslipping, the tips will drop toward the fall line; by bringing the weight to the rear, the heels of the skis will move toward the fall line. This is a means of correcting or controlling the angle of descent during the sideslip (fig. 67).

(4) The speed of descent is controlled by the degree of edging applied to the skis. To stop sideslipping, the edging is gradually increased by pressing the knees forward and toward the slope.

(5) In adverse terrain and snow conditions the aid of both ski poles may be used while sideslipping on the uphill side. The poles are used in the same manner as in pole riding on a traverse. This method adds a third point of suspension and braking action.

116. Step Turn in Motion

a. Use. This method of changing direction while in motion is useful at slow speeds in all snow and terrain conditions. It is particularly useful in adverse snow conditions and in confined areas.

b. Techniques.

(1) Before turning, lead with the ski which corresponds with the direction of turn, i. e., right ski ahead when turning to the right.

(2) In turning to the right the weight is placed upon the left ski, which is then edged to the right. The unweighted right ski is then raised and placed on the snow in the new direction. The weight is transferred to this ski by moving the body in the new direction while pushing off from the left ski. Complete transfer of

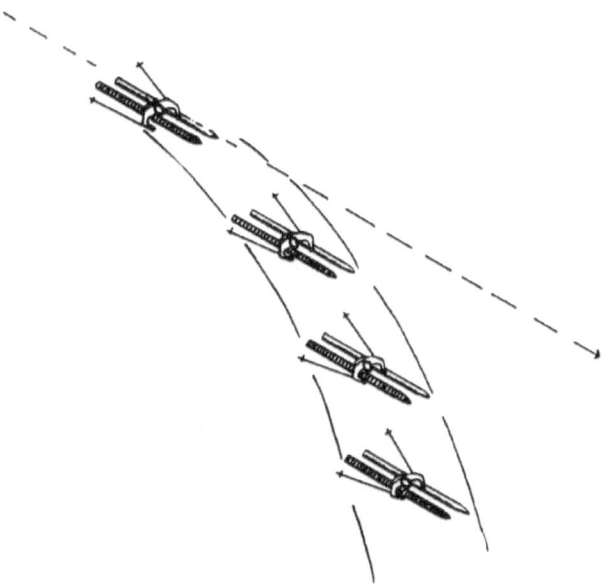

Figure 66. Sideslipping from traverse.

Figure 67. How the body position affects the skis in sideslipping.

body weight is essential. The ski poles are held to the rear (fig. 68). The higher the speed, the more the center of gravity is lowered by bending the knees. This adds stability and aids in keeping up with the turn.

(3) If desired, the steps can be continued as long as the skier is in forward motion and until the desired direction is obtained.

117. Snowplow Turn

a. Use. The snowplow turn is efficient for use at slow speeds, especially when carrying a pack and rifle. Because the snowplow position is retained, this turn enables the individual to maintain good control. In this turn, fundamental body positions and movements are used which are an important part of advanced turns.

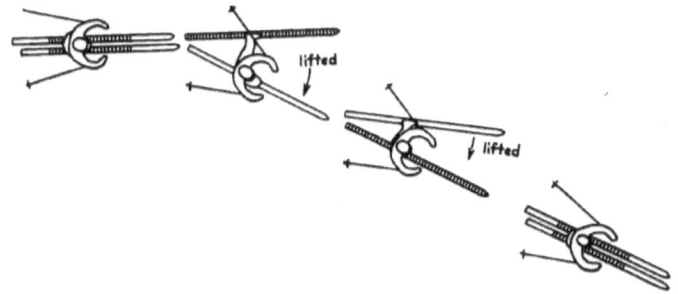

Figure 68. Step turn in motion.

b. *Technique.*
  (1) *Straight down the slope.*
    (a) In executing a snowplow turn to the left while snowplowing directly down a slope, the body weight is transferred smoothly onto and over the right ski by rotating the body and shoulder in that direction. This transfer of body weight initiates the turning action.
    (b) The bend of the right knee is increased forward as the turn progresses. The left knee is kept well bent with this ski flat and unweighted throughout the turn.
    (c) Ski tips remain even and the V angle of the skis constant. Avoid leaning into the slope. Ski poles are carried as in the snowplow position. Care must be exercised to keep them pointed to the rear as the shoulder and body are rotated.
    (d) As the turn is completed the body weight is either placed evenly on both skis to continue in a snowplow or gradually transferred to the left ski to start a turn to the right.
  (2) *From downhill traverse position.*
    (a) In making a turn while traversing, the snowplow position is assumed as described in paragraph 113b(3)(a). The edging of the lower ski is decreased and the body leaned forward in order to bring the skier into the fall line in preparation for the turn, as described above. The turn should be continued till the skier has obtained the desired angle of descent (fig. 69).
    (b) After the snowplow turn has been completed and it is desired to continue with both skis together, as in the downhill traverse, the body weight is kept on the lower ski while the upper unweighted ski is brought parallel with it into a traversing position (fig. 70).
  (3) *Variation.* To make a snowplow turn from a traversing

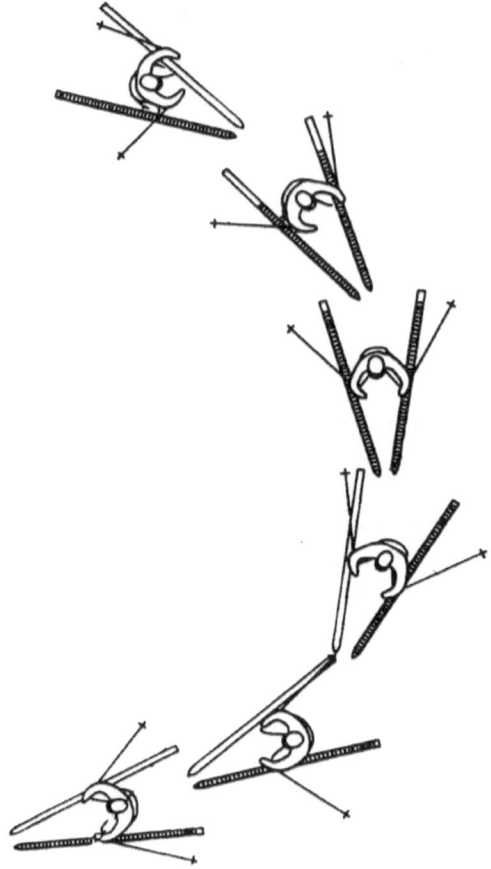

Figure 69. Snowplow turn from traverse.

downhill position in variable snow conditions, and when skiing with a pack, it is advantageous to make the half snowplow with the uphill ski. In this method the body weight remains on the lower ski. The upper, unweighted ski is moved into a half snowplow, kept flat, and the tips of both skis even. The edging of the lower ski is decreased, knees bent more, and the body leaned further forward to bring the skier into the fall line. In reaching the fall line, both skis are brought into a full snowplow and the body weight is gradually shifted over and onto the other ski in executing a snowplow turn as described above.

118. Advanced Turns

The advanced turns used in military skiing are the christiania turns. These are applied at all speeds to change directions, to reduce speed, or to stop. These are the most advanced turns

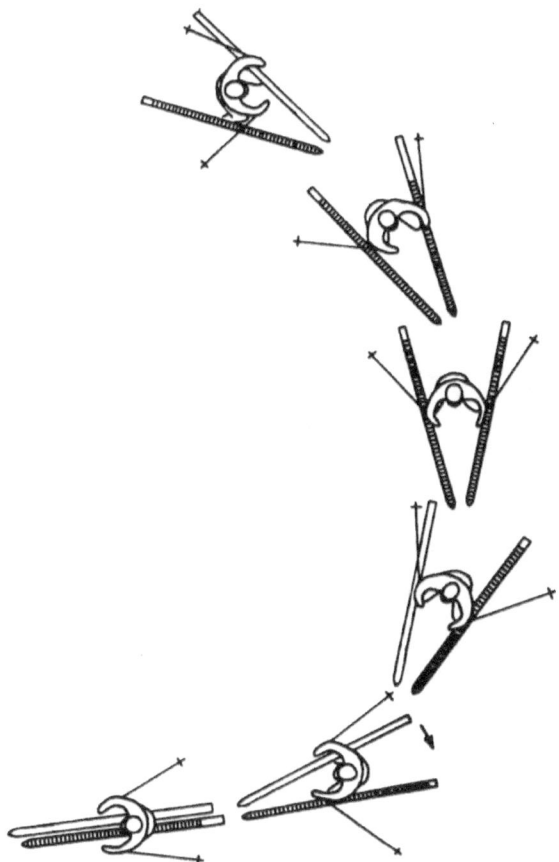

Figure 70. Snowplow turn made from downhill traverse back into traversing position.

taught and are executed with the basic motions already learned, such as forward lean, edge control, and body rotation. The application of these turns may be limited by terrain and snow conditions, as well as the degree of proficiency attained and the load carried by the individual soldier. The christiania turns are started from a variety of positions, but all are completed in the same manner (fig. 71).

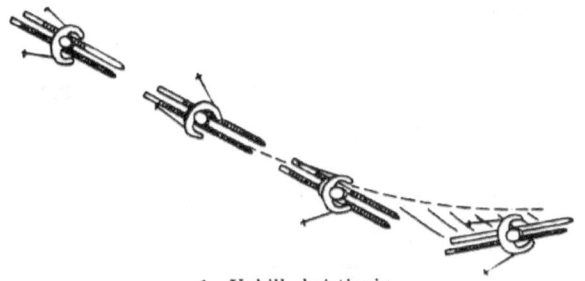

1 Uphill christiania.

Figure 71. The christiania turns.

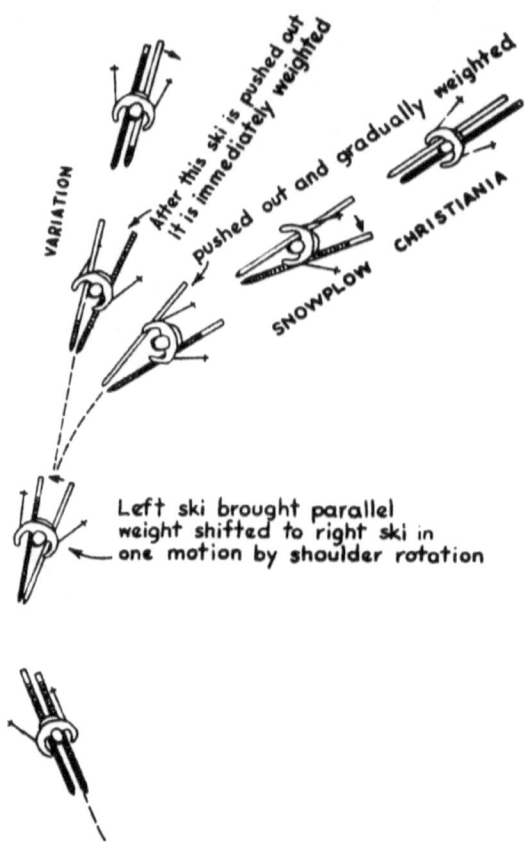

2 Snowplow christiania and variations.
Figure 71—Continued.

119. Uphill Christiania

The uphill christiania is used to turn uphill, to reduce speed and to stop. It also forms the basic movement which is used in completing other christiania turns.

a. In preparing for the uphill christiania during a downhill traverse, the upper shoulder is brought well forward in order to increase the body rotation that will be applied during the turn.

b. The turning action of the skis is started by decreasing the amount of edging and, at the same time, bringing the lower shoulder and hip forward in the direction of the turn in a pronounced rotation. Forward lean of the body and knee bend are increased and the upper ski leads throughout the turn (1, fig. 71).

c. During the turn, both skis are controlled by gradually edging them into the slope. The weight is directly over the skis. Avoid leaning into the slope.

d. Forward lean and body rotation are increased and continued as the turn progresses. Forward speed will gradually decrease, permitting the skier the choice of continuing in a new direction or coming to a stop.

e. Care must be exercised so that the ski poles are not allowed to swing to the front during the rotation of the body.

f. This turn can be made from any angle across the slope to and including the fall line.

g. From a fall line the turn can be made in either direction. In preparation, the ski corresponding with the direction of turn is advanced (i. e., left ski leads for a left turn) and more of the body weight placed on the other ski. Emphasis is given to body rotation and knee bend to initiate the turning action of the skis.

h. To assist the turning action, a down-up motion can be used in this turn. As the turning action is stated as in *b* above, the body is lowered and returned to normal as the turn is completed.

120. Snowplow Christiania

a. Technique. The snowplow christiania is used on turns made downhill while traversing at greater speeds than employed in the basic turns. For this reason the turn looks complicated to the student. Basically, it is a combination of the snowplow turn and the uphill christiania. The basic techniques of the snowplow turn made from a traverse position are also used here to reach the fall line. The uphill christiania is then applied to either change direction or to stop. In combining these methods the speed must be greater, the body weight shifted more rapidly, and the spread of the skis in the snowplow position at a narrower angle. Using the snowplow christiania it is possible to link a number of turns together to control speed in a continuous descent. A breakdown of the technique is as follows:

(1) In making a turn to the left from a downhill traverse, the body weight is shifted momentarily to the uphill ski. The lower ski is then moved downhill into a half snowplow position, keeping the ski tips even. At the same time, the lower shoulder is brought forward. The uphill ski is then pushed uphill to form a snowplow.

(2) Body weight is then transferred back to the upper ski by body rotation, initiating the turning action.

(3) As the fall line is reached, the unweighted left ski is brought slightly forward and parallel with the right ski. The turn is then completed as in the uphill christiania (2, fig. 71).

(4) The upper body is kept from leaning into the slope throughout the turn, especially during the initial turning phase. Forward lean and knee bend are increased. All motions are fluent and smooth and must be well timed during the turn.

(5) When a decrease in speed is desired before starting the turn, there are two methods which can be used. In the first method the lower ski is first placed into a half snowplow position. Temporarily transferring the body weight to this ski and edging it will cause a braking action. When speed has been decreased as desired, the upper ski is pushed upward, the edging of the lower ski decreased, and the turn continued as in (2), (3), and (4) above. In the second method both skis are kept parallel and a sideslip from the moving traverse position is started. Edging of the skis in this movement will provide braking action. When speed has been decreased as desired, the turn is started as from the downhill traverse position.

b. *Variations.*

(1) In difficult snow and terrain conditions another method may be used to execute a snowplow christiania. In making a turn to the left from a downhill traverse with this method, the upper (right) ski is brought into a half snowplow position. Leaning well forward, increasing the knee bend and decreasing the edging of the lower ski will bring the skier smoothly towards the fall line; the body weight is transferred over and onto the right ski in a smooth forward and downward motion, assisted by bringing the right shoulder forward. As the transfer of body weight is completed, the unweighted left ski is brought forward and parallel with the right ski and the turn completed from the fall line as in the uphill christiania.

(2) As more skills and balance are acquired, the snowplow christiania may be done at higher speeds with the angle of turn kept closer to the fall line. In this method only a half snowplow with the upper or lower ski is used in

the preparatory position, or the skis are kept parallel and the fall line is reached with a pronounced knee bend and forward lean of the body while the turn is completed with an uphill christiania.

121. The Lifted Christiania

a. Use. The lifted christiania turn is very useful in adverse snow conditions and in confined terrain where a short radius turn is necessary. It is also useful for skiing at night and with heavy loads, since it is a slow turn made with one ski pole being used to increase lateral stability.

b. Technique.

(1) The turn is started by applying either of the methods described for christiania turns, except that the speed is adjusted to suit the circumstances.

(2) For a turn to the left as the skier approaches the fall line, the left ski pole is placed in the snow forward and down the slope, but not in front of the moving left ski tip. The reach should not be overextended. The right pole is used in the normal manner. Weight is then applied to the ski pole, using it for means of support and as a pivot point.

(3) Body weight is then shifted to the right ski. Since it is difficult to turn the left ski in such a short radius, this ski is lifted and placed parallel to, and slightly ahead of, the right ski, and the turn completed as in the uphill christiania.

(4) This turn can be made without the use of the ski pole. However, the individual should use the pole so that maximum stability and safety will be maintained at all times.

Section IV. MILITARY SNOWSHOEING

122. Purpose and Scope

a. Snowshoes are individual aids for oversnow movement. Like skis, they provide flotation in snow and are useful for cross-country marches and other activities which require movement in snow-covered terrain.

b. The snowshoe is an oval or elongated frame of wood braced with two or three wooden crosspieces and the inclosed space filled with a web of rawhide lacing. A binding or harness attached to the webbing secures the wearer's foot to the snowshoe. Flotation is provided by the webbing, which is closely laced and prevents the snowshoe from sinking too deeply into the snow when weight is placed upon it. Depth and consistency of snow will determine the amount of support obtained on the snow cover and the rate of movement.

c. Snowshoes are particularly useful for individuals working in confined areas such as bivouac sites and supply dumps, for drivers of various types of vehicles, gun crews, cooks, mechanics, and for similar occupations where aids to movement in snow are necessary. Transporting, carrying, and storing snowshoes is relatively easy due to their size and weight. Maintenance requirements are generally negligible and little skill is required to become proficient on snowshoes. However, the requirement for physical conditioning is as great, or greater, as that needed for skiing. The use of snowshoes when pulling and carrying heavy loads is particularly practical, as the hands and arms remain free. On steep slopes, however, the use of snowshoes is considerably limited because traction becomes negligible and the snowshoe will slide, causing loss of footing. Generally, the rate of movement in any type of terrain is slow because snowshoes will not glide over the snow. The gliding properties of the ski are not obtained with the snowshoes; this adversely affects the amount of time and energy spent in movement. In deep snow the trailbreaker must be changed frequently. Especially when wet, snow tends to stick to the webbing, thereby adding weight to the snowshoe.

d. There are two types of standard issue snowshoes, the trail and the bearpaw. Both can be used with all types of winter footgear. The trail snowshoe weighs approximately 6.5 pounds and the bearpaw, 5.5 pounds.

    (1) *Trail.* The trail-type snowshoe is long, with a rather narrow body and upturned toes (fig. 72). The two ends of the frame connect and extend tail-like to the rear. The turned-up toe has a tendency to ride over the snow and other minor obstacles. The excellent flotation provided by its large surface makes the trail snowshoe best for cross-country marches, deep snow conditions, and trailbreaking. The leather binding and toe hole allow the heel to rise freely off the webbing, thus permitting the tail end to drag over the snow and track.

    (2) *Bearpaw.* This type of snowshoe is short, wide, and oval in shape, with no frame extension (fig. 73). The bearpaw snowshoe is preferable to the trail type for close work with weapons and vehicles, in heavy brush, and in other confined areas. Carrying or storing is also easier.

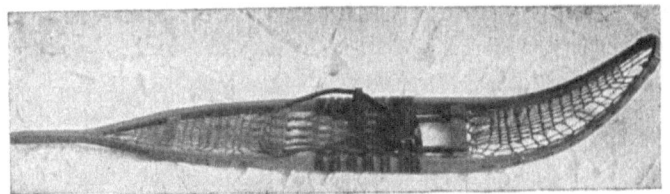

Figure 72. Trail snowshoe.

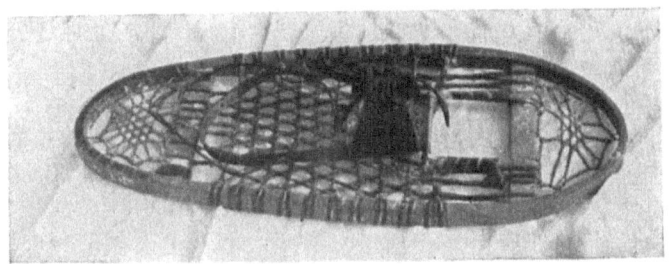

Figure 73. Bearpaw snowshoe.

123. Care and Storage of Snowshoes

a. Care. Snowshoes must always be kept in good condition. Frequent checks are necessary, particularly of webbing and binding, as individual strands may be ripped or worn out. Repairs must be made immediately, otherwise the webbing will loosen and start to unravel. If unvarnished, the rawhide webbing will absorb moisture, stretch and turn white, particularly in wet snow. It should be dried out slowly, avoiding direct flames, and be revarnished at the first opportunity. Wooden frames may fray from hard wear and should be sanded and varnished. When needed, other minor repairs should be made as soon as practicable. When snow cover is shallow, care must be taken not to step on small tree stumps, branches, or other obstacles, since the webbing may be cut or damaged. Stepping into water is to be avoided; the water will freeze and snow will stick to it. When not in use in the field, snowshoes are placed in temporary racks, hung in trees, or placed upright in the snow. They should be kept away from open fires and out of reach of rodents.

b. Storage. In off-seasons, snowshoes are stored in a dry, well ventilated place so that the rawhide will not mildew or rot and the frames warp. Each snowshoe is closely checked for possible damage, repaired if needed, and revarnished. As in the field, snowshoes are protected against damage and from rodents.

124. Snowshoe Technique

a. A striding technique is used for movement with snowshoes.

In taking a stride, the toe of the snowshoe is lifted upward to clear the snow and thrust forward. Energy is conserved by lifting it no higher than is necessary to clear the snow and slide the tail over it. If the front of the snowshoe catches, the foot is pulled back to free it and then lifted before proceeding with the stride. The best and least fatiguing method in travel is a loose-kneed rocking gait in a normal rhythmic stride. Care is taken not to step on or catch the other snowshoe.

b. On gentle slopes, ascent is made by climbing straight upward. Traction is generally very poor on hard-packed or crusty snow. Steeper terrain is ascended by traversing and packing a trail similar to a shelf across it. When climbing, the snowshoe is placed as horizontally as possible in the snow. On hard snow, the snowshoe is placed flat on the surface with the toe of the upper one diagonally uphill to get more traction. In the event the snow is sufficiently hard-frozen to support the weight of a person, it is generally better to remove the snowshoes and proceed temporarily on foot. In turning around, the best method is to swing the leg up and turn in the new direction, as in making a kick turn on skis.

c. Step over or around obstacles such as logs, tree stumps, ditches, and small streams. Care must be taken not to place too much strain on the snowshoe ends by bridging a gap, since the frame may break. In shallow snow there is danger of catching and tearing the webbing on tree stumps or snags which are only

Figure 74. Making a kick turn on snowshoes.

lightly covered. Wet snow will frequently ball up under the feet, interfering with comfortable walking. This snow should be knocked off with a stick or pole as soon as possible. Although ski poles are generally not used in snowshoeing, one or two poles are desirable when carrying heavy loads, especially in mountainous terrain. The bindings must not be fastened too tightly or circulation will be cut off, and frostbite may occur. Leather straps stretch easily. During halts, bindings should be checked for fit and possible readjustment.

125. Training

Snowshoe training requires little technical skill. However, emphasis must be placed on the physical conditioning of the individual and the development of muscles which are seldom used in ordinary marching. The technique, as such, can be learned in a few periods of instruction. Stiffness and soreness of muscles are to be expected at first. The initial training should be gradual with regard to loads carried and distances covered. It should be progressive, with ample time allowed for the individual to acquire physical proficiency, gradually increasing the distance covered and weight carried or pulled. Overcoming obstacles such as dense brush, fallen timber, and ditches should be emphasized during training. Trailbreaking, with frequent change of the lead man, should also be stressed. Snowshoe training can be accomplished concurrently with other training requiring individual cross-country movement.

Section V. APPLICATION OF SKI AND SNOWSHOE TECHNIQUE

126. Skiing in Variable Terrain and Snow

a. General. As a military skier the individual must be prepared to move in a great variety of terrain and snow conditions during daylight and darkness. He must be constantly alert in order to judge conditions on the route ahead and to offset the sudden changes often encountered. The techniques of skiing which he has learned will allow him to operate effectively on slopes only if he is capable of applying these methods properly and of keeping his skis under control at all times.

b. Variable Terrain. The forward lean of the body must be increased as a slope suddenly steepens, since skis will slide faster. The opposite is true as the slope is lessened. Generally, the body should be nearly perpendicular to the slope to insure proper balance. When skiing over bumpy terrain, the stability of the skier is greatly disturbed. To minimize this the knees are kept supple to act as shock absorbers, permitting the center of the body to maintain as straight a line as possible. To further increase stability on large bumps the skier increases knee bend, lowering the body when approaching the top of the bump, riding over it in this position, and then assuming a normal running position as soon as the top is passed (fig. 75). This action will lessen the chance of the skier being thrown into the air. When moving through a hollow the normal ski and body position is maintained, with the knees absorbing the sudden change of pressure. In deep snow the leading ski should be further advanced to improve balance. Under those conditions where cross-country skis and bindings are used, or it is necessary to ski downhill without the use of the downhill hitches to secure the heel, the center of gravity must be kept lower by more bending of the knees. As forward lean of the body is not practical under these conditions, weight shift will need to be controlled to a greater extent by the knees and the advancement of one ski in front of the other.

c. Variable Snow. When skiing from soft snow onto hard snow the forward lean of the body must be increased, since the skis will gain speed and have a tendency to run from under the skier. The opposite is true when running from hard snow onto soft snow (fig. 76). In this case the body leans slightly to the rear and the leading ski is advanced farther ahead just before the soft snow is entered. Lateral stability can be increased by extending the arms sideways as is done when attempting to

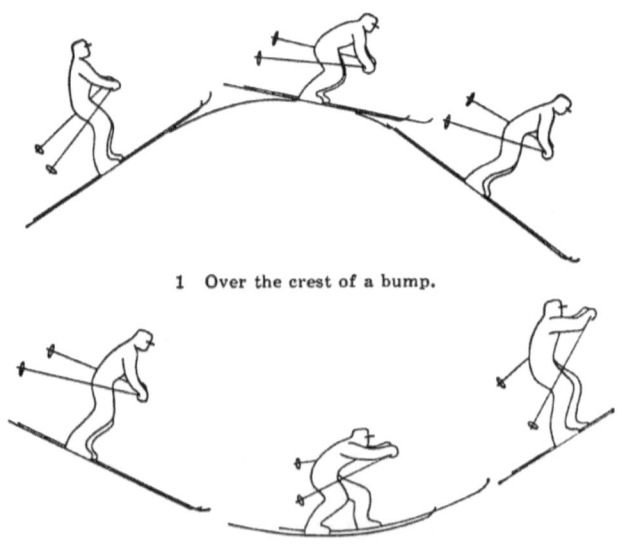

1 Over the crest of a bump.

2 Through the hollow of a bump.

Figure 75. Position of the body in downhill bump riding.

walk a railroad track, but the ski poles must still be kept pointing to the rear. When skiing on icy crust, stability is improved by keeping the skis farther apart or by running in a slight snowplow position. However, if the slope is rutted snowplowing may become hazardous because the tips tend to get caught. To control speed under these conditions, sideslipping and pole riding may be used. Pole riding is less effective and in extreme cases the use of sideslipping may become necessary. On icy snow the skis may chatter in a turn. To correct this, body weight is kept well forward and the edging of the skis carefully controlled as the turn is made. Crusty snow which will not support the skier's weight (breakable crust) is the most difficult to cope with. Speed is kept slower while making all turns. It may become necessary to use the step turn in motion or a kick turn to change direction.

d. Forests. Due to the limited skiing room in wooded terrain, movements for changing direction must be rapid and of shorter radius than in open terrain, especially during downhill movement. In addition, the skier must be more alert so that obstacles may be quickly overcome with a minimum of delay. The step turn in motion is a very useful technique for changing direction in this type of terrain, but speed must be reduced to use this technique. In descending narrow trails in wooded terrain or during night movements, the half snowplow or pole riding are useful for control of speed. During unit movement in wooded terrain, one man falling can block the progress of all personnel behind him. If

Figure 76. Running from hard packed snow onto soft snow.

an individual falls he should remove himself from the track in the fastest way possible, even if this results in losing his original position in the column. The baskets of ski poles have a tendency to snag branches during movement in wooded terrain, resulting in loss of balance. To avoid this as much as possible, the shafts of the ski poles should be pointed directly to the rear.

127. Obstacles

a. General. Snow-covered terrain will contain many small obstacles such as fences, tree windfalls, and small streams or ditches. The individual must be skilled enough to cross them easily to save time and energy. Crossing obstacles can be very time consuming for a unit. Wherever possible, the men should be dispersed so as to enable them to cross on a broad front. In some cases the overall time needed can be reduced if skis are removed while overcoming the obstacles.

b. Fences and Windfalls. Low fences and windfalls (one to two feet high) are crossed by skiing or snowshoeing beside the obstacle so that the skis or snowshoes are parallel and alongside it, then stepping over first with one foot then the other. If obstacles are high and cannot be crossed underneath, the skis or snowshoes are placed alongside and a kick turn made over the top. In the case of rail fences or large diameter windfalls it may

Figure 77. Crossing obstacles.

sometimes be easier to sit on the obstacle and swing both feet simultaneously to the other side. High barbed-wire fences can be crossed by removing pack and rifle and crawling underneath (fig. 77).

c. Ditches or Small Streams. These are crossed by stepping over them sideways, using the ski poles for support (fig. 77). If the ditches are deep and wide it is better to descend to the bottom either by sidestepping or sideslipping and then climb the other side by sidestepping. However, care must be taken to avoid rocks or other obstacles which might damage the skis or snowshoes.

d. Steep Slopes. When it is necessary for troops to descend or ascend slopes which are too steep for their ability, or where traversing is not practical, the skis should be removed and the slope negotiated on foot whenever snow depth will permit.

128. Skiing with Pack and Weapon

a. General. When skiing with pack and weapon, the most practical technique for moving over rolling terrain is the one step. However, terrain and snow conditions will make it necessary to continually vary the techniques applied. A pack and weapon do not necessitate a change in techniques, but the heavier load changes the center of gravity and will affect the manner in which various movements are made.

b. Effects on Movements.
   (1) Lunges are shorter and pushes with poles less powerful.
   (2) To aid in maintaining balance when skiing downhill through rough terrain, the leading ski is advanced farther and the knees kept more flexible than when skiing without a load.
   (3) Speed of descent is reduced and techniques are applied more cautiously.
   (4) Rotation of arms and shoulders is made with less vigor and emphasis.
   (5) Slopes are climbed with a more gradual traverse.
   (6) When skiing through woods or in brushy terrain, care must be exercised in order to prevent any protruding parts of the weapon from catching on branches, causing loss of balance.
   (7) In the event of a fall it is sometimes more efficient to remove the pack and weapon before attempting to regain footing.

129. Sled Pulling

a. General. Pulling a sled is hard work, but it will be easier if the proper techniques are used. The movements and techniques used should be within the ability of all members of the team. Generally speaking, the methods of hauling sleds apply to both skiers and snowshoers.

b. Preparation for Sled Pulling.

   (1) The tow ropes must be of the proper length and also properly laid out and fastened by snap buckles in tandem system (fig. 78). The sled harnesses are adjusted to fit loosely on the individuals.

   (2) If skis are to be used for pulling, they must be properly waxed. More emphasis must be placed on insuring good holding capacity of the wax on the snow. However, sliding capacity should not be entirely forfeited.

   (3) Proper loading and lashing of sled must be checked before moving out.

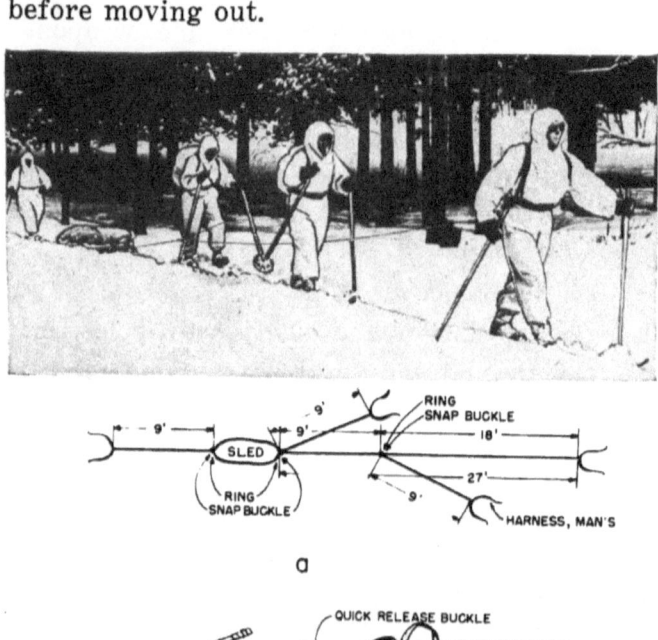

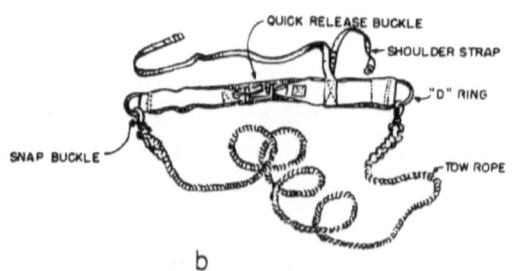

a. Pulling Arrangement.
b. Harness, Man's Sled.

Figure 78. Modified tandem system.

c. Pulling in Varied Terrain. When pulling a sled over comparatively flat terrain, skiers normally use the one step ski technique. When crossing small ditches, the sled is stopped in the ditch while the pullers go as far as the tow ropes allow. Then, by a simultaneous pull, the sled is brought up out of the ditch. To change direction in woods, the pullers continue to move straight forward until the sled comes to the desired turning point. The pullers then move in the new direction with the turn being controlled by the puller nearest the sled, assisted, if necessary, by the man behind, who lifts the sled from the rear. When the forest is dense and space does not allow the men to move far enough ahead before the turn is made, the pullers must start the turn by gradually making as gentle a curve as possible while the two men nearest the sled (in front and behind) guide, lift, and otherwise assist in turning the sled. While turning, the pullers must watch the movements of each other in order to avoid confusion.

d. Uphill Climbing. To pull a sled uphill the following methods can be applied:

   (1) On short, gentle slopes the herringbone can be used.

   (2) On a steep, short slope the pullers can use the sidestep (fig. 79). In this case the rear man moves to the side of the sled and, while sidestepping, assists in pulling the sled by using the rope fastened to the rear end.

   (3) If the slope is comparatively long, the uphill traverse is employed. Ski climbers can be used if the length of the slope justifies the time required to put them on.

   (4) In difficult terrain a relaying technique may be used when the necessary equipment is available. In this technique a climbing rope, nylon, 120 feet long, or similar item, is fastened to the sled. The pullers then climb uphill as far as the rope allows. Standing in place, the sled is then pulled up to their position. This

Figure 79. Sled team using the sidestep technique for ascending a steep slope.

procedure is repeated as many times as is necessary to reach the top. When using this technique care must be taken to insure that the sled is well anchored each time the pullers move up, since a run-away sled may not only damage itself but is a serious hazard to anyone below. Where steep slopes must be ascended for considerable distances, less energy will be expended if the sleds are left behind and the sled load backpacked to the objective.

 e. Downhill Movement. In descending a slope the following methods can be used:

(1) On short, gentle slopes skiers can take a sled down by using a double poling technique or one step. However, it will be necessary to maintain sufficient speed to prevent the sled from overrunning the pullers. The rear man can assist in this by braking the sled, although in most cases very little braking will be needed. If the team is on snowshoes, the pullers can descend normally while the man in the rear insures that the sled does not overrun those in front.

(2) A short, steep slope can be descended by sidestepping either on skis or snowshoes. The rear man must do all the braking. Skiers can also use sideslipping for this type of terrain. On short descents in wooded terrain such as river banks, the rear man can keep his ski poles in one hand and grasp at trees with the other hand to assist in letting the sled down slowly.

(3) On long, moderate slopes skiers can use the snowplow as a braking method (fig. 80). If more braking is necessary than can be supplied by the rear man, the puller closest to the sled may move to one side or he may remove his rope and refasten it to the rear of the sled and assist the rear man for more effective braking, if space

Figure 80. Sled team descending a slope using the snowplow technique for braking.

allows. Snowshoers on this type of slope may also change pullers to brakers to aid in descent.

(4) On a long, steep slope requiring the team to go straight down, all men will be needed to brake the sled. This can be done by fastening all tow ropes to the rear of the sled with all men braking from the rear. The snowplow or sideslipping techniques are used as the braking method.

(5) Traversing by both skiers and snowshoers may be used on long, steep downhill slopes. In this case the puller nearest the sled and the rear man should remain above the sled and as far from it as the ropes will allow. From this position they can brake, preventing the sled from sideslipping.

(6) In very steep terrain a long rope, when available, may be used to lower the sled straight down the slope. This procedure is the reverse of the uphill relay method described in $d(4)$ above and is a very practical method for evacuating wounded.

130. Skijoring

a. General. Skijoring, as used in this manual, is the term applied to moving men on skis over snow by towing them with vehicles. This provides a faster and less tiring method for individual movement than is possible under their own locomotion. Oversnow vehicles, track and wheeled vehicles can be used for pulling skiers (fig. 81). The best routes for skijoring are snow covered roads and trails, frozen lakes, rivers, or paths made by tracked vehicles. Speeds up to 15 miles per hour may be maintained on level ground by trained troops, depending on weather and trail conditions. Normally, one rifle squad can be towed behind a light carrier and two squads behind a squad carrier. Towing more than two squads by one vehicle is impractical, due to the delay caused by the increased number of individuals frequently falling.

b. Use of Tow Ropes.

(1) Two ropes 120 feet long are used for towing a rifle squad behind a vehicle for the purpose of securing sufficient space between the individuals. The skiers, in columns of twos, are spaced at equal intervals behind the vehicle and outside the ropes. A gap of approximately 13 feet is left between individuals.

(2) Several methods of towing can be used according to the situation, the terrain, and the distance of movement:

(*a*) Each skier ties and uses a short butterfly knot in the

Figure 81. Skijoring.

        rope, holding his poles in his outside hand (1, fig. 82). For making a butterfly knot see FM 31–72.
- (*b*) Each skier ties a long butterfly knot (using 5 to 6 feet of rope). He places the loop of the rope around the buttocks and leans against the loop. He does not go inside the butterfly knot (2, fig. 82).
- (*c*) Using the ski pole method (3, fig. 82), the skier rests both arms and body and can arrive at the destination in better physical condition. Another advantage in this method is that a skier can easily exercise his hands to prevent frostbite during movement in extreme cold.
- (*d*) When being towed through dense wooded areas, or when contact with the enemy is imminent, skiers may simply grasp the rope without tying the butterfly knot or using ski poles as a rest. Thus, they can maneuver through narrow trails and are more ready for immediate combat.

  (3) No matter what method of towing is being used, individuals must never be allowed to fasten themselves to the tow rope. In case of a fall they must be able to release their hold immediately to avoid serious injury to themselves or other skiers. The ski poles are usually kept in the hand and available for instant use. During training and in combat situations when contact with the enemy is not probable, the ski poles may be loaded on the vehicles to avoid accidents.

  c. Skijoring Technique.
    (1) The track is made as simple as the terrain permits.

Steep slopes, obstacles, and sharp turns are avoided. When these cannot be bypassed the speed must be reduced in order that the skiers can maneuver. A high degree of cooperation between the driver of the towing vehicle and the skiers is necessary. One man, usually the assistant driver, is responsible for stopping or slowing the vehicle in order to prevent casualties due to speed or obstacles. He constantly observes the skiers and other vehicles, gives the driver orders, and signals the skiers when the vehicle will slow down, speed up, or stop.

(2) When the vehicle begins its forward movement each man on the rope should move forward under his own power for a few steps, gradually placing tension on the towing rope to prevent being suddenly jerked into motion, causing a fall. When under way, the skier's body is leaned slightly backward, the knees are bent slightly, and the upper body is nearly straight. Skis may be farther apart than in normal skiing. One ski is kept slightly ahead. The position should be one in which the skier can relax but still be alert to sidestep quickly in order to avoid obstacles and maintain his balance. If a skier falls, he should release the towing rope immediately.

(3) When approaching a sharp curve where the area for movement is confined, the vehicle should be slowed down or, in some instances, stopped. The skiers on the inside of the curve simultaneously move the rope overhead to the inside until the curve is negotiated. Failure to do this may result in being pulled off balance by the sudden change in direction as the vehicle completes the turn.

(4) When descending hills the men can brake by using the snowplow or half snowplow, if space allows, to prevent overrunning the vehicle or, if conditions warrant it, they may move to the side of the track where the softer snow will decrease their speed. If the terrain will not allow for controlled braking and collision with the vehicle seems imminent, the individuals should release the rope and disperse to the sides of the track. On short downhill slopes the vehicles should increase speed temporarily so that the men need not brake. On long, steep slopes the men can descend independently of the vehicle and reattach themselves after the slope has been negotiated.

1   Short butterfly knot

2   Long butterfly knot

3   Using ski poles as a rest

Figure 82. Methods of towing or skijoring.

CHAPTER 5
MOVEMENT

Section I. PROBLEMS AFFECTING MOVEMENT

131. General

The lack of roads, the soft, wet terrain prevalent in the summer, the snow and blizzards in winter, thick forests in mountains and hills, and the innumerable waterways are some of the barriers to movement in most cold areas of the world. The ability to overcome the many obstacles to movement may well be the deciding factor in winning or losing a war in these cold areas. In the large, undeveloped cold areas of the world the possibility for maneuver is limited mainly by the ability to move. Mobility begins with the individual.

132. Influence of Seasonal Changes in Weather and Terrain on Mobility

a. Spring Breakup and Fall Freeze-Up.

(1) The spring breakup and fall freeze-up periods are by far the most difficult seasons in which to maintain mobility. The period of breakup may last from 3 to 6 weeks and will present restrictions to movement (fig. 83). The snow becomes slush and will support little weight. Winter roads break down, the ice in waterways melts, rivers are swollen and become torrents. Movement at this time of year poses many problems, however, movement is possible in cold areas at all times. Normally, at this time of year, temperatures drop at night, freezing the surface, and mobility during this period can be maintained.

(2) The period of freeze-up with rain and open or half-frozen waterways will also present barriers to movement. Complete freeze-up may take up to 3 months, often restricting the movement of heavy equipment across lakes until late January.

(3) The early winter period, when there is little snow and the ground and waterways are firmly frozen, will provide excellent trafficability for foot soldiers and vehicles.

Figure 83. Breakup season.

   b. Winter. The low temperatures, snow, blustery winds, and bulky clothing and equipment required during winter hinder movement as it is known in more temperate climates. By the proper use of specialized equipment for cold weather operations, mobility can be maintained. Using skis, snowshoes, oversnow vehicles, and aircraft, mobility is possible. In the barren tundra or on ice caps the hard snow found in these areas will readily support an individual on foot as well as oversnow vehicles. In the forested areas the snow will normally be deeper and the temperatures lower. The depth of the snow and the trees in these areas will prove to be the greatest obstacles to mobility. With the proper equipment, however, such as skis and snowshoes, mobility may be maintained.

   c. Summer. The open, uninhabited, and roadless Far North creates certain restrictions on the rate of movement in summer. The permanently frozen subsoil of the Far North is called permafrost. The topsoil may thaw as a result of the sun's rays during the long summer days. The terrain varies from hard, rocky, raised spines of land to the flat marshy areas of swamps. The vegetation is mostly stunted. Scrub willows, birch, alder or mountain ash, grass, moss, and lichen are prevalent. The northern terrain is a mixture of shallow sloughs, swamps, rivers, and lakes. Cross-country travel on foot is tiring but possible. Wet, spongy ground and brush will slow down foot movement. Selection of routes is extremely important to avoid unnecessary bogging down of the main body.

   d. Forested Areas. A great portion of the North is covered with evergreen forests and with numerous swamps and water courses. Few trails exist through the forests and those that do exist are of poor construction, making progress difficult and slow. In situations such as this, full use should be made of all waterways, as they are the best means of ground travel in the winter.

Section II. FOOT MOVEMENT

133. General

Both summer and winter cross-country travel in the North is difficult, each with its own complexities. Of necessity, travel will be slower. However, with the proper training in the use and maintenance of equipment, the proper enthusiastic leadership, and the will to accomplish the mission, nothing is impossible.

134. Basic Rules for Foot Movement

The following guides are based on experience factors and should be considered in preparing for cross-country movements in the northern areas.

a. Insure that all personnel participating in the move are fully aware of the mission, route, etc. Equipment must be checked and loads evenly distributed. Dispatch trailbreaking teams far enough in advance to insure continuous, uninterrupted movement of the main body. Men should be dressed as lightly as possible consistent with the weather to reduce excessive perspiring and subsequent chilling. Complete cold weather uniforms must be available while operating in cold environments. A large proportion of cold weather casualties result from too few clothes being available to individuals at such time as a severe change in the weather occurs.

b. The first halt after initiating a march should be made in approximately 15 minutes. This will allow adjustment of clothing and equipment. Subsequent halts should be frequent and of short duration to insure rest and to prevent chilling. Halts should, so far as possible, be made in sheltered places which will provide protection from the elements. Warm drinks should be provided during the march if possible.

c. The buddy system is mandatory in the North and men must be instructed to watch their buddy carefully in extreme cold for early signs of frostbite. Individuals must not be allowed to fall out of the line of march, except in an extreme emergency. If this should occur, proper care must be taken to insure that he does not become a cold weather casualty. Normally, the second-in-command will bring up the rear of the column and, at each halt, will check the men and report their condition to the leader.

d. Prior detailed reconnaissance is most important to insure successful mobility in the northern areas. Maps of the area are not, in all cases, accurate. Streams overflow their banks and snow piles deep in drifts. Consequently, the only sure way to avoid long, back-breaking detours for the marching element is to give maximum attention to prior reconnaissance.

e. Marching in single file is often the best formation. It maintains track discipline, camouflage, and reduces the number of trailbreakers and reconnaissance parties required. Natural obstacles may limit the use of other formations. In large groups, columns in single file are excessively long and susceptible to ambush from the flanks or rear. Tactical considerations may make other formations more suitable. When following vehicles, the double tracks may be used as a pathway.

Section III. TRAILBREAKING

135. General

a. Purpose.

(1) The purpose of trailbreaking is to make the march of the main body as easy and fast as possible in order that the troops will arrive at their destination in good fighting condition. Trailbreaking accomplished at any time of the day or night through deep snow and difficult terrain is hard and time-consuming work. The progress of trailbreaking is dependent on the terrain, weather and snow conditions, vegetation, physical condition of the trailbreaking detachment and, finally, on the tactical situation. Therefore, plans must be carefully made and trailbreaking parties well organized.

(2) In addition to trailbreaking, the mission of providing frontal security for the main body is a normal function of the trailbreaking party. In order to maintain the integrity of the command, approximately one-fourth of the unit is given the mission of trailbreaking and security of the march. For example: The battle group on axis of advance normally assigns one rifle company to lead (trailbreaking). The quartering party may accompany the trailbreaking party or may follow later. The company in turn assigns one rifle platoon to lead, functioning simultaneously as a trailbreaking party for the lead company. Since the trailbreaking unit is the first to arrive in the new bivouac area, its commander is also responsible for establishing temporary security of the area. When the quartering party arrives in the bivouac area they will perform the normal functions of a quartering party as outlined in FM 101-5.

b. Planning. Based upon the estimation of the tactical situation, terrain, weather and snow conditions, the most suitable route is selected for the movement. As a general rule terrain features which offer least resistance will be followed. The following principles are guides for selection of the route:

(1) *Open terrain.* In order to keep the main body sufficiently dispersed, ski tracks are more widely separated in open terrain than in covered terrain. For concealment, normally only one ski track is broken over open fields, or the track is broken close to the edge of the forest so that shadows may help to conceal it. All individuals must follow this track. Breaking trail through dense brush is a time-consuming job and should be avoided when possible. In open terrain the oversnow vehicles should be used for breaking trail and for towing the trailbreaking party by skijoring to the maximum extent to save time and energy of the individuals. This system is never used in brush or rugged terrain, because twigs and brush overrun by the vehicle stick up from the trail and become entangled with the ski bindings, making skijoring very difficult. Skijoring over rugged terrain is not practical unless a well broken trail is available.

(2) *Covered terrain.* Whenever possible, time and situation permitting, the trail should follow along forest terrain with little or no underbrush. It provides good concealment and protection against wind. The trail should be broken close to bushy trees in order to provide better concealment. Thickets and windfall forest areas should be avoided, as it requires a great amount of effort to break a trail in areas of this type. If a triple track is broken for sleds, wide curves must be made when changing direction and the bushes and branches must be cut from the inside of the curve. The thoroughness with which the small trees, bushes, and branches on both sides of the broken track are cleared will depend on the time allowed the trailbreaking party.

(3) *Hilly and mountainous terrain.* When the situation permits, the trail is broken along the valley. Rivers frequently afford the easiest route in this type of terrain. If the valleys cannot be used, the trail should be broken on the lee side of the hill if it does not make the track too much longer than that of the straight course. Use gentle inclines when climbing uphill or descending. When tracks are broken downhill the speed of the trailbreaking party is

often slow, due to soft and deep snow. However, when packed, the same tracks may make the speed of the skiers in the main body too fast. This will result in many falls, especially during darkness.

(4) *Water routes.* Frozen lakes, rivers, and creeks offer the most suitable routes for the trails. They also help in land navigation. For best protection and concealment, the trail-breaking party skis very close to the shore or on the bank, as this facilitates better concealment of the individuals and units, their trail, and any quick movements into the wooded areas of the shore. Sometimes in winter, and especially in the spring, there may be water under the snow surface on the lakes and rivers, thus causing the running surfaces of the skis to freeze. Check for concealed water under the snow before starting to break trail across the ice. Areas in which water is found under snow should be bypassed. If this is not possible, the crossing site must be reinforced with snow or with a combination of brush and snow. Also, the thickness of the ice must be carefully checked before using any ice route. The minimum thickness of ice for one rifleman on skis is one-and-a-half inches; for an infantry column in single file on foot, 4 inches; and for the single light artillery piece or ¼-ton truck, 4 x 4, 6 inches. See load bearing capacity tables in FM 31-71.

(5) *Obstacles.* Since even minor obstacles retard the march, they are bypassed whenever possible. If a wide obstacle is met, such as a ridge or a steep river bank, several sets of tracks are broken over the obstacle so that the main body can cross it on a broad front. Trees and brush are cut well below the bottom of ski tracks in order to avoid twigs and branches entangling in ski bindings and tow ropes. Obstructions such as fences may be cut in order to allow the skier to pass through.

(6) *Weather and snow conditions.* In early winter there is more snow in open terrain than in dense forest; therefore, the trail should be broken close to the forest edge. In late winter the reverse is true. In early spring more snow can be found in ditches, ravines, and on the shadowy side of hills. Forests give good protection against the wind.

(7) *Darkness.* Skiing and snowshoeing at night is slow and exhausting. Therefore, the trail for a night march must be broken along the easiest terrain available. Avoid all rough terrain if possible. Navigation of the trailbreaking

party demands special skill in darkness and during a heavy snowfall. Rivers, creeks, ridge lines, and forest boundaries should be used as aids to navigation in spite of the fact that the broken trail might become longer. Due to the darkness it may be necessary to leave guides posted at locations where the main body may take the wrong course.

(8) *Enemy activity.*
   (a) When breaking trail within the frontline area, the requirements for concealment are most important. Therefore, the trailbreaking party is forced to ski along covered terrain whenever possible. However, if the mission requires fast movement, a trail is broken along the shortest course, paying less attention to concealment. Normally, the mission of security for the march of the main body is given to the trailbreaking party.
   (b) These responsibilities affect the course of the trail. The track should be broken close to such terrain features as hills, forest edges, and lake shores, which facilitate observation. Also, the route should follow terrain which offers a sound approach and suitable places for temporary defense. Sometimes it is necessary to check critical terrain features located near the trail before the trailbreaking party moves forward. Elements of the trailbreaking party may occupy certain security positions and remain stationary until the main body has passed these critical points, at which time they may rejoin the rear of the column. For the purpose of deceiving the enemy, it may be desirable to create numerous false trails criss-crossing and angling off in all directions. In burned-over areas or thin deciduous forests, concealment from aerial observation is practically impossible. A single track clearly indicates the whereabouts and approximate size of the unit making it. Miscellaneous trails, therefore, create confusion and optical illusions.

(9) *Number of trails used.* The number of trails to be broken depends upon the size of the column using them, the tactical situation, and time available for trailbreaking. An organization of battle group size normally requires two or more march tracks and one or more communication tracks for messenger service and control of the march column. In cases where time is very limited for preparations, only one track may be established for a battle group. When contact with the enemy becomes imminent, greater emphasis is placed on security and less emphasis placed

on trailbreaking. The possibility for a rapid deployment of the troops requires that the number of trails or tracks be increased from that of a routine cross-country march.

c. Organization. The trailbreaking party preceding units mounted on skis must also be mounted on skis. The trailbreakers of elements on snowshoes are also mounted on snowshoes. Mixing of skiers and snowshoers on the same track is not recommended. Snowshoes tend to compact the snow on ski trails making it difficult for the main body to follow on skis.

(1) The lead company will normally be assigned the mission of breaking trail for one complete day. It is replaced by another company on the following morning. One rifle platoon at a time is assigned as lead platoon and is called a Trailbreaking Party. It may also include engineers whose duties would include reconnoitering ice routes, seeking suitable terrain for permanent type winter roads, preparing ice reinforcements, and performing other engineer tasks. Forward observers may also accompany the trailbreaking party.

(2) Depending on terrain conditions, 1 to 2 oversnow vehicles are assigned to the party to be used for breaking trail in open terrain, skijoring, and carrying individual manpacks and platoon equipment. In unfavorable terrain conditions the vehicles remain under company control or with the higher echelon. In either case the trailbreaking party will place their individual manpacks in the unit vehicles prior to their departure. The trailbreaking party consists of its organic rifle squads, called Trailbreaking Squads. A trailbreaking party is expected to break trail approximately a half a day at a time, but may be rotated sooner depending on local conditions. Trailbreaking squads, in turn, are normally rotated as often as necessary in order to maintain the speed necessary to complete the mission in time.

d. Trailbreaking Squad. The organization, duties, and special equipment of the trailbreaking squad are indicated in figure 84. Squad leaders must insure that their men have a sufficient number of tools of proper size before moving out. The tools are part of the tent group equipment and are used in preference to intrenching tools. To conserve energy and to assure an uninterrupted march, the leading man (breaker) of the squad is regularly relieved. In very deep and heavy snow a relief may become necessary every 200 yards. When the change is ordered by the team leader, the man to be relieved steps sideways out of the path and falls in at the rear of the team. The man following him then becomes the breaker. Special equipment is exchanged by passing it to the next man in line during the rotation. The breaking team will be relieved by the reserve team as directed by the squad leader whenever the point team tends to slow down due to fatigue.

e. Trailbreaking Party. The trailbreaking party consists of two or more trailbreaking squads. Normally a rifle platoon will be assigned this mission, especially if the snow is heavy and the weather severe.

(1) One of the squads is always designated as the base squad and is responsible for navigation and the general direction to be followed. The platoon leader and the navigation detail directly under his control will follow the base squad. When dead reckoning is required, the base squad breaks the center trail and works slightly ahead of the other squads for the purpose of maintaining the proper direction of the

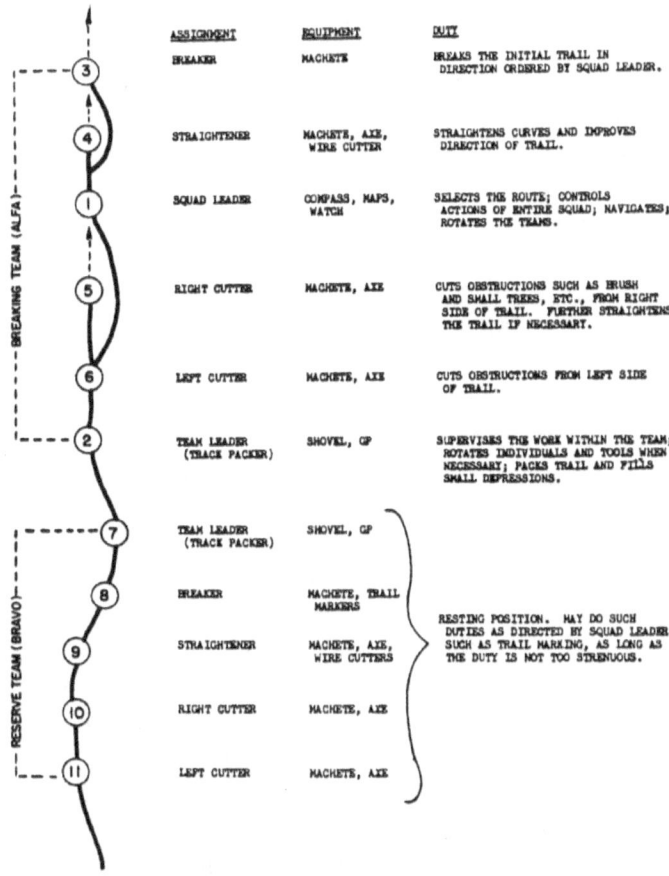

Figure 84. Organization of trailbreaking squad.

squads which are moving on both sides of the track made by the base squad (fig. 85). In cases where the party follows easily recognizable terrain features, such as small creeks or the edge of open terrain, the base squad follows next to this terrain feature, making navigation easier. The other squads are echeloned to the right or left, and their breaker (the first man) to the right or left of the last man of the squad ahead (fig. 86).

(2) Interval between the tracks varies from about 15 yards in covered terrain to approximately 100 yards in open areas, depending on the local situation. The depth of the party varies from 100 to 200 yards. Members of the weapons squad (its MG's and RL's being carried in unit vehicles or sleds) may be assigned to the navigation detail, to flank security missions, to assist the vehicles in breaking their trail off the ski tracks, and similar duties. The weapons squad may follow and improve the trails being established, as directed by the leader of the trailbreaking party. From the area where vehicles are temporarily halted due to the close proximity of the enemy, one track may be widened into a triple track to facilitate the movement of heavy weapons, ammunition, and warming tents. This equipment is usually moved forward by man-drawn sleds.

(3) The trailbreaking party moves far enough ahead of the column to permit a steady rate of march by the main body. This distance varies according to the situation, snow, and weather conditions, and terrain encountered. As a rule of thumb, the trailbreaking party precedes the main body by 1 hour for each 3 miles of marching distance. For example, if a 15-mile march is planned, the trailbreakers leave 5 hours in advance of the parent unit.

f. Techniques. The trailbreaking squad may break a normal or triple track as required. On *normal track* the first man makes his tracks so that the grooves are a little wider apart than usual, approximately 1 foot. The trailbreaker usually uses the one step technique. In deep and soft snow, however, his steps will be shorter than normal and he will be forced to lift his skis at each step to prevent the tips from running under the surface of the snow. Progress will be slow and may be exhausting. Therefore, the man in the breaker position must be rotated often.

(1) When track-laying vehicles and cargo sleds cannot be used any further due to the tactical situation, the crew-served weapons, ammunition and warming tents must be moved to the units in man-drawn sleds. Therefore a triple track

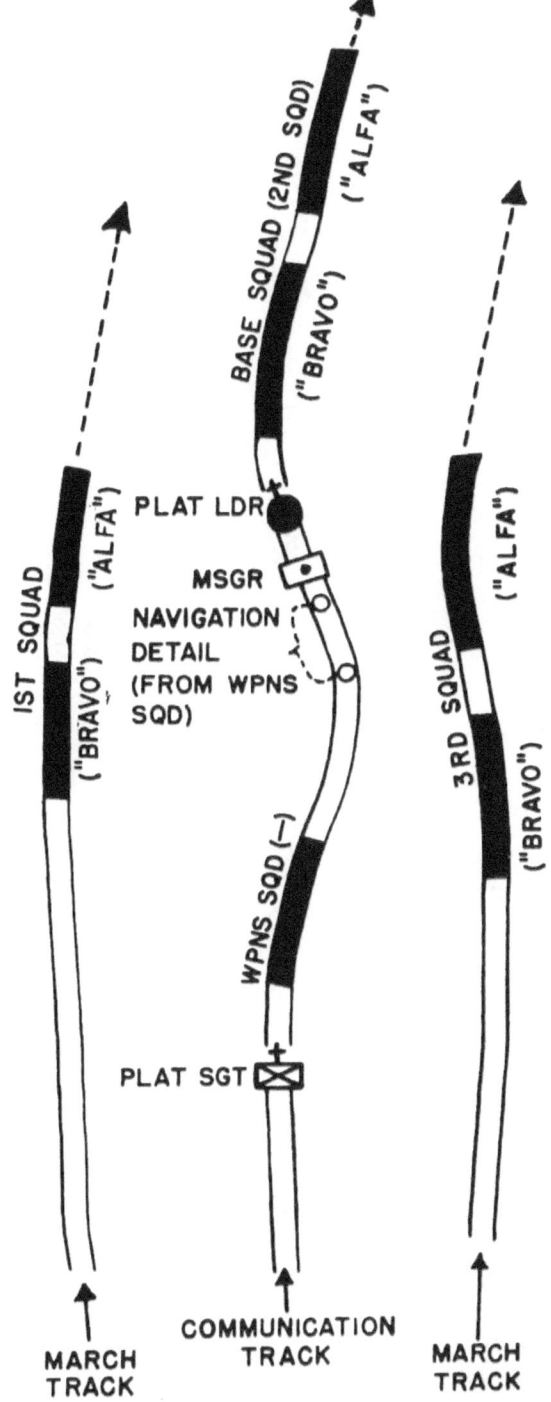

Figure 85. Trailbreaking party (dead reckoning).

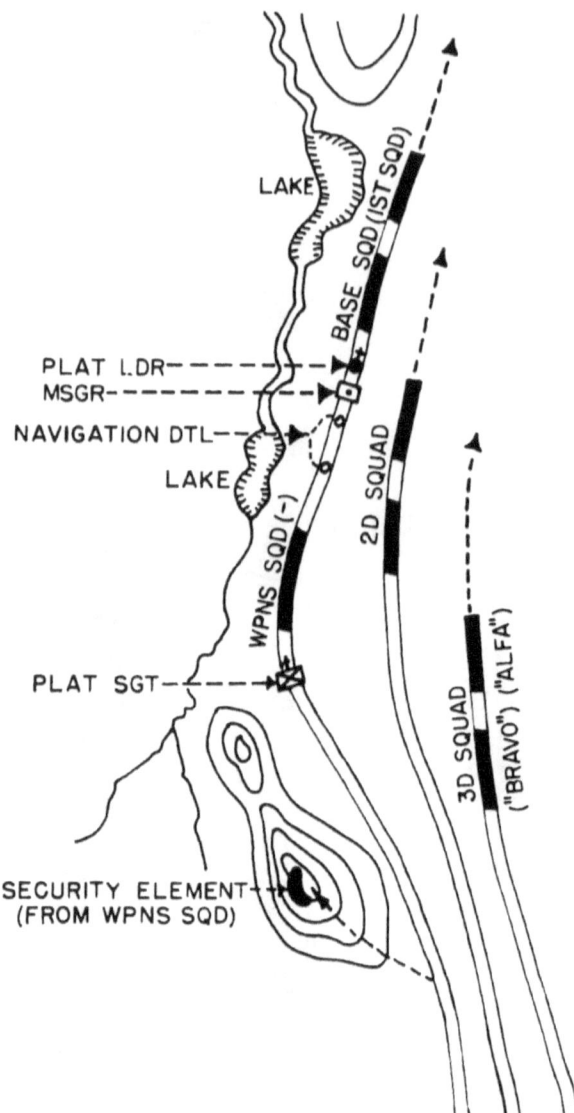

Figure 86. Trailbreaking party following recognizable terrain features.

is broken because the normal track is too narrow. When starting a triple track (1, fig. 87), the leading three men of the breaking team will break a normal track of two grooves. The third groove is started by the fourth man who keeps one ski in the already broken groove and makes a new groove with his left (right) ski, depending on which side of the original groove the new groove will be broken. Alternate men behind the fourth man, both in breaking and reserve teams, ski along the original grooves made by the first three leading men, the others following the

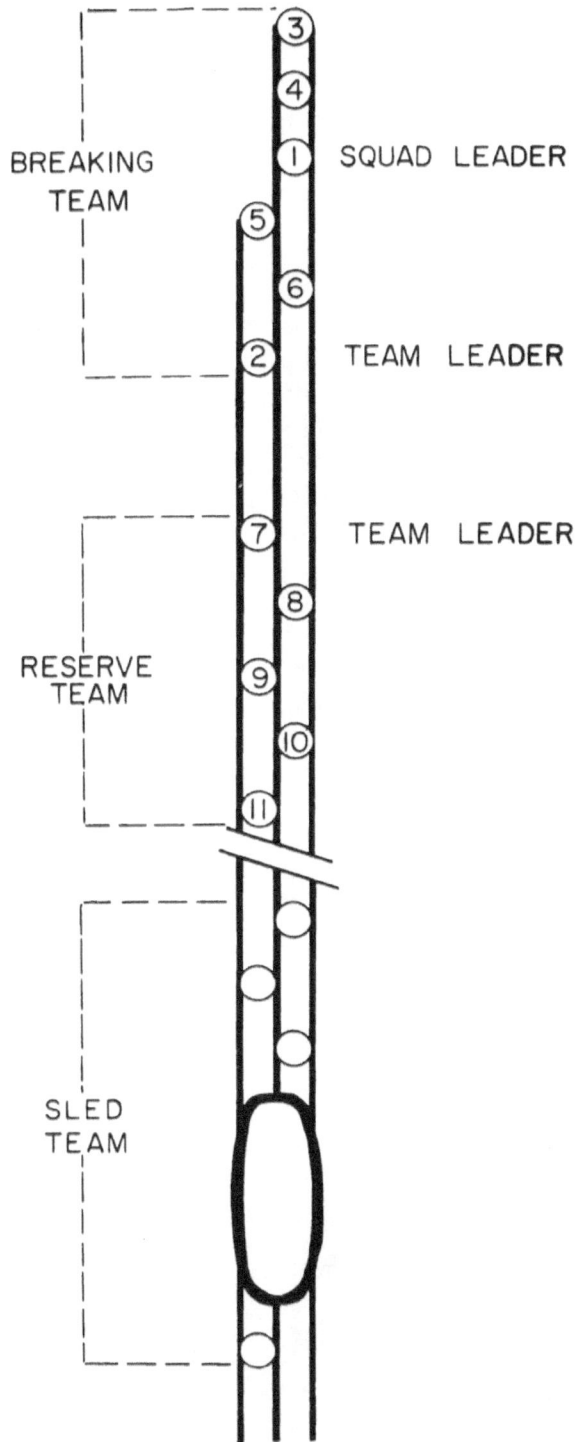

1 Organization of the trailbreaking squad.

Figure 87. Breaking of Triple Track.

grooves made by the fourth man. This creates a trail with three grooves, a triple track (1, fig. 87). This provides the proper type of track for pulling man-drawn sleds. Due to the fact that sleds tend to destroy the ski tracks, only one of the ski trails will be prepared as a triple track and this track will be used for man-drawn sleds only.

(2) Ski tracks must be kept separate from the trails and roads established for vehicles and cargo sleds, due to the fact that the vehicles tend to destroy the ski tracks and, conversely, the skiers on the winter road tend to harass the vehicular traffic. Signal wire layed alongside the ski track must be located far enough to the side so as not to become entangled with skis and ski poles. When crossing the ski track the wire must be buried well below the track or secured overhead, whichever is most desirable.

136. Marking the Trails

a. The trailbreaking squad marks its trails as uniformly as possible. The types of markings used must be known to the unit that follows. When several squads are operating, marking by the base squad is usually sufficient. The marking is simple, and recognizable by night as well as by day. Temporary trails through new snow need

2 Triple track completed.

Figure 87—Continued.

simple markings only where the tracks or roads are crossed by other trails. Trails that are frequently used for long periods are more permanently marked. The following can be used as trail markers:

(1) Twigs on trees and shrubs broken in a predetermined manner, or nicks in tree trunks made by using a hatchet or machete.

(2) Poles or guiding arrows planted in the snow.

(3) Markers made of rags or colored paper.

(4) Trail markers (willow wands).

(5) Rock cairns or piles.

b. Snowfalls, fog, poor observation, and uniformity of the terrain necessitate thorough and frequent markers spaced at uniform intervals and numbered successively in the direction of march. To avoid the destruction of trail markers by traffic, the markers are placed about 3 feet off the trail. When strange tracks cross the trail of the unit they are obliterated at the point of crossing. Sentries are posted at crossings, if necessary, to direct units that follow.

Section IV. LAND NAVIGATION

137. Effects of Environment

a. General. Basically, map reading, as well as navigation under cold weather conditions, follows the same principals as in the temperate zones. In addition to the normal procedures, every individual must be well familiar with certain conditions peculiar to the cold weather regions and the techniques applicable to navigation. Due to the fact that a technical failure or human error may easily, and especially in the winter, be fatal to the individual or to a unit, great care must be exercised when navigating in low temperatures.

b. Navigation Problems. The following conditions, characteristic of the cold weather regions, will make accurate navigation very difficult:

(1) Lack of adequate large scale maps in the sparsely populated areas which will increase the requirements for and the use of aerial photographs.

(2) Photos of many areas will be difficult to read and interpret because of the monotony of detail, absence of relief and contrast, and absence of manmade works for use as reference points.

(3) Dense forests and wildernesses offer few landmarks and limit visibility. Also, barren, monotonous tundra areas

north of the tree line are characterized by lack of landmarks as aids for navigation.

(4) In winter, short daylight, fogs, snowfall, blizzards, drifting snow, especially in the barren areas, drastically limit visibility. At times an overcast sky and snow-covered terrain create a condition of visibility (whiteout) which makes recognition of irregularities in terrain extremely difficult.

(5) Heavy snow may completely obliterate existing tracks, trails, outlines of small lakes, and similar landmarks. Because the appearance of the terrain is quite different in winter from that in summer, particular attention must be paid to identifying landmarks, both on the ground and in aerial photos.

(6) Magnetic disturbances are encountered, making magnetic compass readings difficult and sometimes unreliable.

(7) Magnetic declination in different localities varies considerably, and must be taken into consideration when transposing from a map to a compass.

(8) Handling maps, compass, and other navigation instruments in low temperatures with bare hands is difficult. Removing handgear may often be possible for a very short period of time only.

138. Methods of Land Navigation

a. The normal methods of land navigation under cold weather conditions remain the same as anywhere else. Maps and aerial photos may be used alone during daylight in terrain which offers enough distinctive terrain features to serve as useful landmarks. They may also be used in conjunction with a compass, especially in terrain which contains insufficient landmarks or under circumstances when visibility is limited.

b. Depending on various conditions, certain supplementary methods, such as position of the sun in daytime, North Star and Big Dipper at night, as described in FM 21–26, may be used to aid in land navigation. Where possible, these methods should be employed in conjunction with the normal methods described above.

c. It is obvious that on vast barren grounds as well as in wide forest, navigation by dead reckoning often becomes the only practical method. Dead reckoning is the process by which position at any instant is found by applying to the last determined position the direction and distance of the course traveled. This method should also be used in areas where landmarks are very limited or totally nonexistent. It is also desirable when the landmarks are obliterated by the limited visibility.

139. Navigation by Dead Reckoning

Navigation by dead reckoning is performed in accordance with FM 21-26. Due to the peculiarities of the cold weather regions, the following hints should be observed when applicable:

a. Responsibility for navigation is assigned to a detail of one officer or noncommissioned officer and 1 to 2 men, all thoroughly experienced in navigation techniques. The detail is placed directly under the control of the unit commander and must be released from the carrying of individual heavy loads and from details such as trailbreaking in order to perform their duties properly. Using a small detail rather than a single navigator is based upon the fact that the method of pacing distances in deep snow has to be modified as described in *c* below.

b. In general, the navigation detail is responsible for—
  (1) Accumulating necessary instruments and equipment.
  (2) Keeping instruments and equipment serviceable.
  (3) Performing the detailed duties of taking and recording necessary data for precise location at all times.
  (4) Maintaining liaison with the commander of the unit.
  (5) Supplying data to keep the column on course.

c. Due to the sliding capacity of the skis, normal pacing system is very inaccurate or, in certain cases, such as on steep slopes, entirely useless. Pacing on snowshoes can be done in emergency. It must be borne in mind, however, that an individual mounted on snowshoes takes much shorter paces than on foot. The only recommended method for accurate ground measurements is a piece of line or field wire (preferably 50 yards long) used by two navigators.

d. Keeping a log is mandatory. The preparation of the log as well as plotting the route from the log data on the face of the map or on a separate piece of paper at the same scale as the map must be completed prior to the departure, to minimize the use of instruments and equipment in low temperatures with bare hands.

e. Certain mechanized aids are highly valuable for navigation by dead reckoning:
  (1) A magnetic compass has been developed for mounting in all vehicles except armored.
  (2) The use of gyroscopic instruments in military vehicles has never been entirely successful, because the require-

ments of such instruments have been extremely difficult to assemble into the manufactured item.

(3) Odograph M1 is an instrument which automatically plots the course of a moving vehicle. It consists of three principal units—the compass, the plotting unit, and the power pack. All components are interconnected by electric cable and flexible shafts. It was originally designed for use in the 1/4-ton truck, but can be used in other vehicles to include tracklaying vehicles and sleds for operation under winter conditions.

(4) Odograph M2 is much more accurate and convenient to use than the M1. It utilizes the miniature gyrocompass for the input of direction. In normal operations, if the map coordinates of the starting point are set on the instrument, it will provide the true coordinates of any point along the course of travel.

(5) The use of rotary wing aircraft for "pathfinding" in bush country greatly assists in land navigation. From the tactical point of view, however, it is less feasible because it tends to disclose the movement. If this method is used, tactical situation permitting, the aircraft may land every 30 minutes and the pilot or navigator mark the correct direction by colored smoke grenades. The smoke, set off near the aircraft while the blades are revolving, tends to disappear quickly but gives time enough for azimuth reading by the navigation detail with the march column.

Section V. ACTION WHEN LOST

140. General

Prior march reconnaissance includes memorizing details of the country to be traversed. Routes should be plotted and as many landmarks located as possible to insure that personnel will not be without recognizable features for any appreciable length of time. If on barren terrain, all navigation instruments must be thoroughly checked and one of the most experienced men should be given the job of navigating and maintaining the "dead reckoning log." It is possible to become temporarily lost while operating in friendly areas or enemy terrain, as on a long range patrol. Each situation should be considered separately, and the main point to remember in any case is to remain calm.

141. When Lost Within Known Locality

If the sector is quiet and there is an absence of war noises or aircraft to guide the patrol toward friendly lines, stop in place.

In a wooded area steps should be retraced to the last known point. If this is not practical, estimate the present location and send a small detail in search of the next known point. Opinions should be taken from the group as a whole if it is felt they will contribute. Search parties must mark their trail carefully in order that they may return and guide the main group forward or rejoin the group should their search be fruitless. In the meantime, the remainder of the group should seek shelter. If it is still not possible to locate the route, carry out the group action discussed in paragraph 142.

142. Conduct When Lost

At the first suspicion that a patrol or unit is not on the right course, it should not keep moving in the hope that it will come across a known landmark. The leader should halt the patrol, not cause unnecessary panic by appearing concerned, and immediately make a detailed check of the route starting at the last known point passed. If extensive checking of the position does not clarify the situation, inform all concerned personnel of the circumstances. When it has been determined the group is definitely lost, the patrol leader must accomplish the following:

a. Seek a shelter, evaluate the situation, and formulate a plan.

b. Gather all food and drink and institute a rationing system.

c. Send a few selected personnel to search for a route, while the balance of the party remains in a sheltered position.

d. Arrange necessary ground-to-air signals (app. III).

Section VI. MECHANIZED AID TO MOVEMENT

143. Track-Laying Vehicles

a. General. So far as small units and individuals are concerned, vehicles of the track laying type are the best aid to movement in northern regions. Deep snow and extreme cold impose special problems of operations and maintenance (app. II). Mandatory characteristics of any vehicle to be used in support of small units and individuals in the Far North during all *seasons* are mobility over muskeg and tundra, through brush and light timber, and the ability to break trail in deep snow. A complete discussion of these problems is beyond the scope of this manual. This manual is limited to a brief discussion of the general capabilities and employment of vehicles which are capable of tactical cross-country movement during all seasons. In order to conserve the energy of troops, mechanized transportation of heavy weapons, ammunition, tentage, sleeping equipment, rations, and individual packs must be utilized to the maximum. Troops burdened with carrying or pulling these items soon become exhausted and lose their mobility and fighting capac-

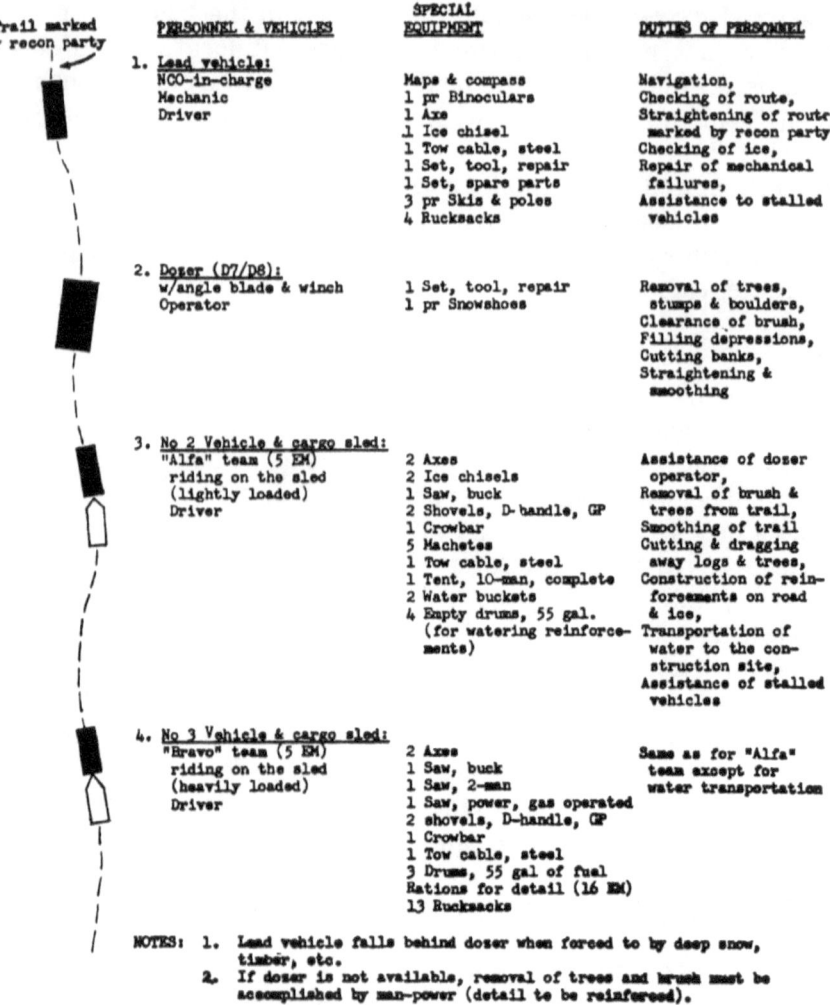

Figure 88. Construction detail for winter roads.

ity. Wheeled vehicles are generally restricted to road movements and have little use in cross-country operations of small units. The series of pictures contained in figures 88 through 94 illustrate construction problems entailed in negotiating winter trails with track laying vehicles.

b. Tractor Trains. The purpose of tractor trains is to furnish oversnow movement of supplies and equipment. Tractor trains will be utilized normally from a railhead, truckhead, or airhead to the division or battle group supply point. The tractor train is a means of moving large quantities of supplies cross country. The trains are composed of cargo sleds or wanigans drawn by construction-type tractors and normally, due to their size and slow rate of march, are not used forward of the battle group supply point. The tractor

Figure 89. Road reinforced with snow.

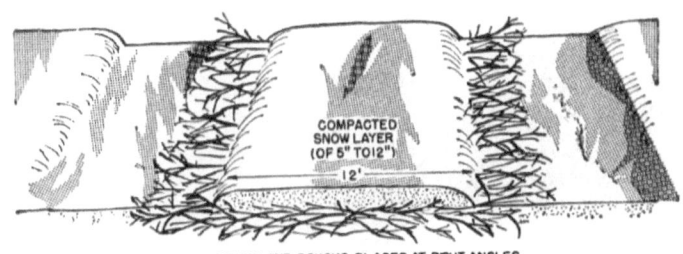

Figure 90. Road reinforced with brush and snow.

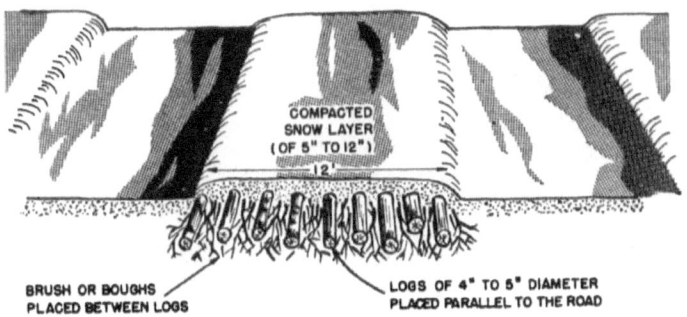

Figure 91. Road reinforced with brush, logs, and snow.

train in no way takes the place of wheeled cargo carriers that may be able to operate on roads or trails.

c. Types of Oversnow Vehicles. All tracklaying vehicles have the same general cross-country characteristics, although they vary in size, gross weight, and carrying capacities.

144. The Squad Carrier

The full track cross-country vehicle presently in use is the Armored Personnel Carrier M59, designed for cross-country operations. This combat vehicle may be used for multiple purposes and is considered to be the best vehicle presently available for use by

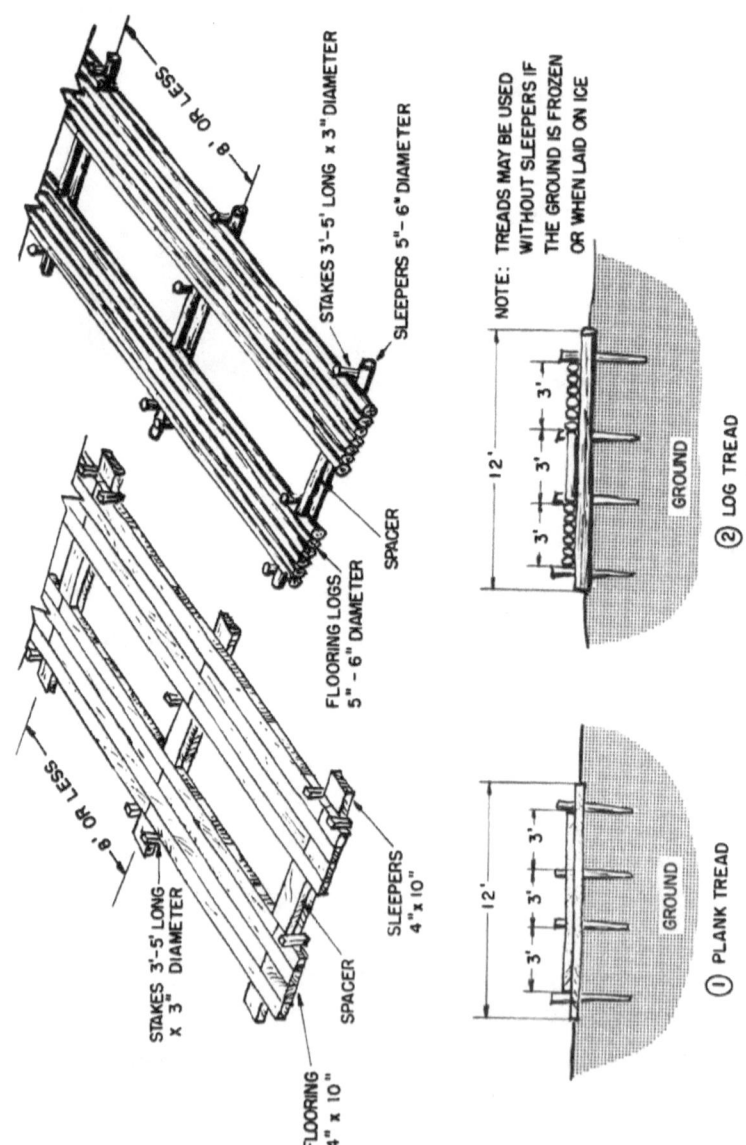

Figure 92. Road reinforced with tread.

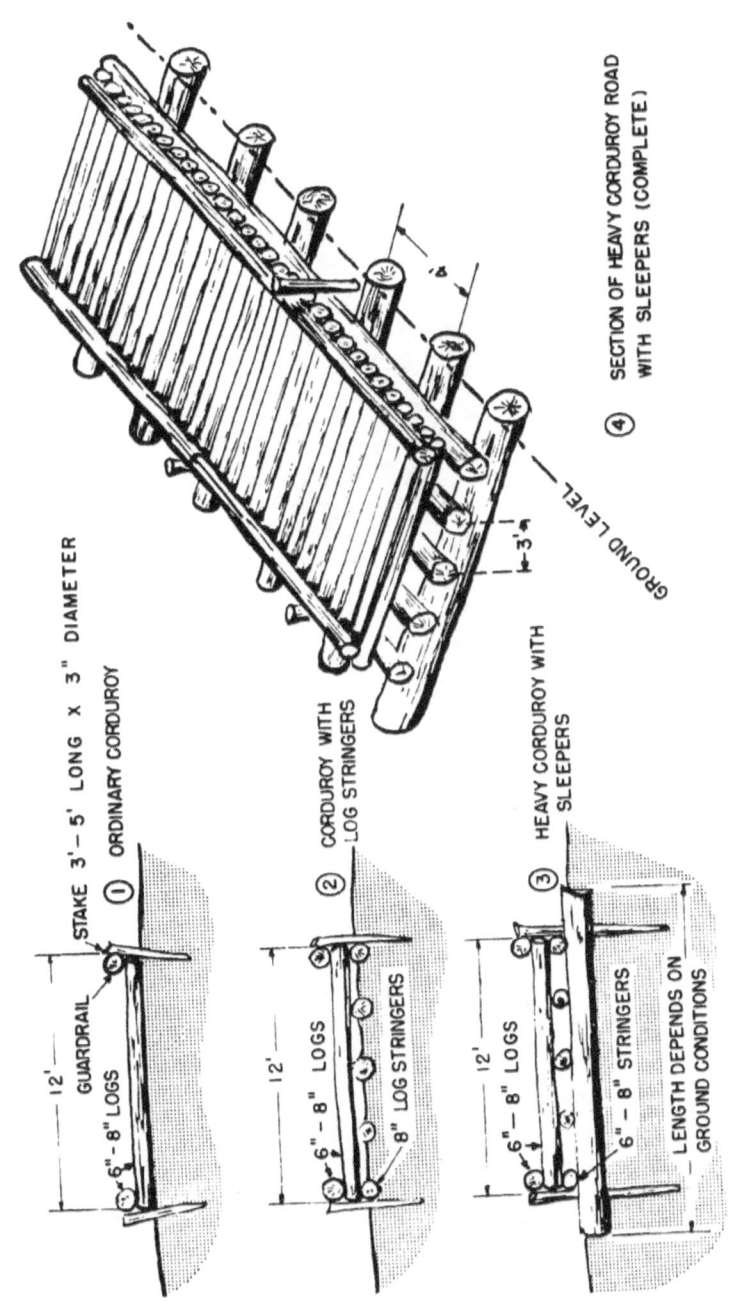

Figure 93. Types of corduroy roads.

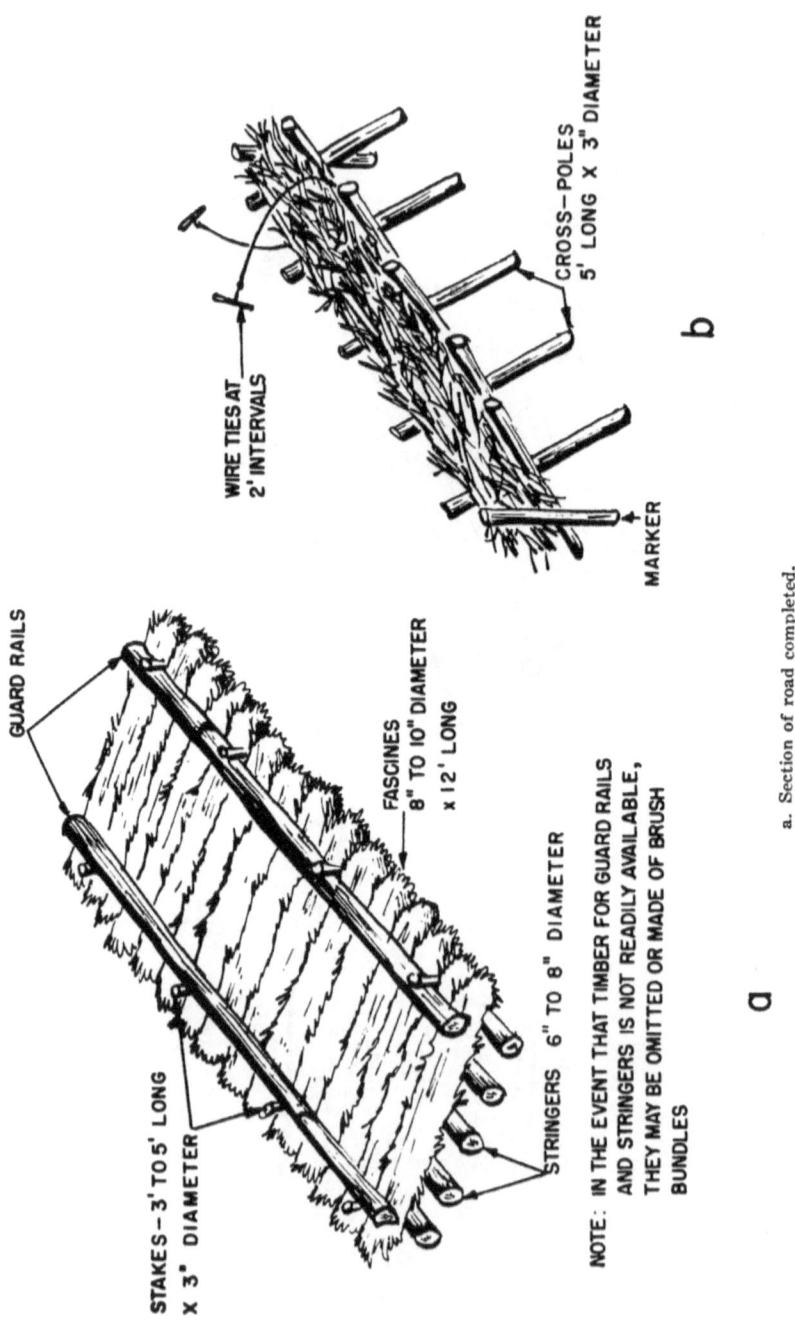

a. Section of road completed.
b. Use of "Sawhorse" for making brush bundles.
Figure 94. Corduroy fascine road.

Figure 95. The squad carrier.

the combat troops in the North. It combines many of the features found to be desirable in the North. As a squad carrier it transports a full rifle squad together with their equipment and impedimenta. In an emergency it can furnish heat and shelter and sleeping accommodations for the squad. It can function as a cargo and weapons carrier, a reconnaissance vehicle, or a command post. For evacuation purposes it transports five litter patients. Armament consists of one caliber .50 machine gun mounted on the turret. The completely inclosed and watertight hull provides an amphibious capability and some protection against radioactive fallout. For additional data on the current model (fig. 95), see TM 9-2300-203-12.

145. Tanks

Tanks are designed for cross-country mobility to include traveling in deep snow. In addition to their normal tactical missions they may be employed to transport personnel in an approach march and, in an emergency, to tow skiers. Windchill factors must be taken into consideration prior to moving troops on tanks for any appreciable distance to insure against frostbite. Tanks may also be used to pull cargo sleds; however, damage can be caused to sled tongues by the fast, jerky starting which is a characteristic of the tank. Tank tracks provide excellent routes of advance for troops, especially in the assault phase of the attack.

Section VII. SLEDS

146. Man-Hauled Sleds

a. Sled, Boat-Type, Plastic. Man-hauled sleds are necessarily light. They can carry a load of 100 to 200 pounds over difficult terrain and are used for carrying tents, stoves, fuel, rations, and other necessary items of each tent group. They are also used for carrying weapons and ammunition. They may be used as a firing platform for machine guns in deep snow and are particularly useful in the evacuation of casualties. Sleds are seldom used by small reconnaissance patrols because of the decreased speed of the individuals. Strong combat patrols, however, frequently use them for carrying their equipment or for evacuation in cases when faster means are not available. Sleds are provided with white canvas covers for camouflage, to hold the contents in place and protect them from the elements.

   (1) The sled has an approximate weight of 38 pounds, is 88 inches long, 25 inches wide, and has a depth of 8 inches. It is towed by a team of four men. For the purpose of towing, a harness, sled, single trace, is provided. It con-

sists of a loose-fitting web belt which is fastened at the side by a quick release buckle, an adjustable shoulder strap which supports the belt at the desired position on the hips, and a nine foot towing rope with snap buckles at each end. Metal D-rings are position at the front and rear of the belt.

(2) Normally, sleds are towed by manpower during the last phase of an approach march or a similar type movement. In mountainous terrain, when the grades are too steep for vehicles, the equipment must be moved by means of aircraft or manpower. Under other circumstances, however, the equipment may be carried on cargo sleds towed by tracked vehicles. A number of loaded sleds, boat-type, can be placed in cargo sleds (1 ton or heavier) or, in an emergency, can be hooked on improvised tow bars and towed behind the tracked vehicles. A triangle made of green poles and attached to the rear of the vehicle or cargo sled provides an excellent "tow-bar." Four small sleds can be tied to the bar, side by side.

(3) The sled, because of its boatlike shape, is easily maneuverable under a variety of snow and terrain conditions. It is superior to flat surfaced toboggans in maneuvering over difficult terrain, especially in deep snow and in heavily wooded areas.

(4) It is important to distribute the load of the sled properly (fig. 96). In loading, place heavy equipment on the bottom and slightly to the rear and lighter equipment toward the top, in order to prevent the loaded sled from being top heavy. After the sled has been loaded, the canvas covers of the sled should be folded over the load. To keep snow from getting under the canvas and to keep the load from shifting, lash the load tightly by criss-crossing the lashing rope from the lashing ring or from one side of the sled to the other. Place tools such as shovels, axes, and saws on top of the load outside the canvas so that they are readily available for trailbreaking and similar purposes during the movement.

b. Improvised Sleds. Different types of sleds can be improvised from skis, plywood, lumber, or metal sheeting (fig. 97).

147. Cargo Sled and Wanigans

a. For military purposes sleds are classified light or heavy. Light sleds are under 5-ton payload capacity, and sleds with payload capacity of 5 tons or over are considered heavy.

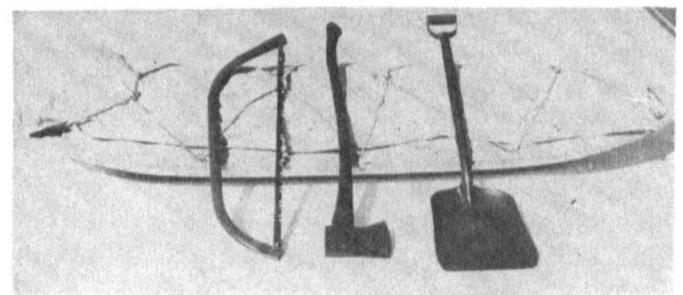

Figure 96. Sled, boat-type, plastic with tools.

Figure 97. Improvised ski sled.

b. Light sleds presently in use are designed to carry 1- or 2-ton payloads. The 1-ton cargo sled (fig. 98) is normally used with the light carrier as a prime mover; and 2-ton sleds (available in limited quantities but not a standard item) with the squad carrier or tractor as a prime mover. Light sleds are suitable for use when rapid travel is involved and in areas where the freezing season has mean temperatures which do not form more than moderate thicknesses of ice on rivers and lakes.

c. Heavy sleds (of a commercial type) which may be used are of 10- to 20-ton payload capacity (fig. 99). It is anticipated that the bulk of supply will be transported on heavy sleds as opposed to

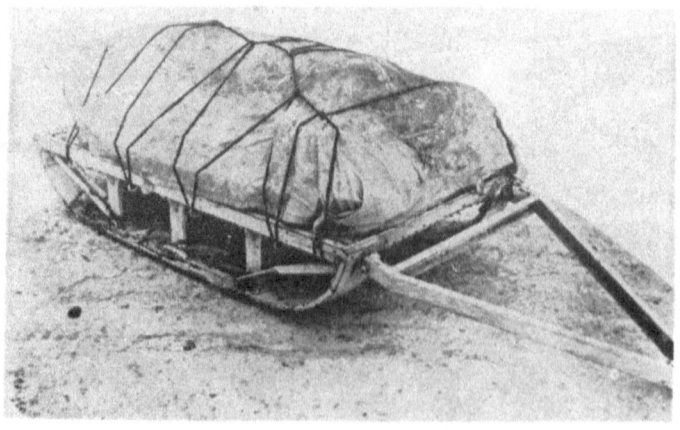

Figure 98. Sled, Cargo, 1-Ton.

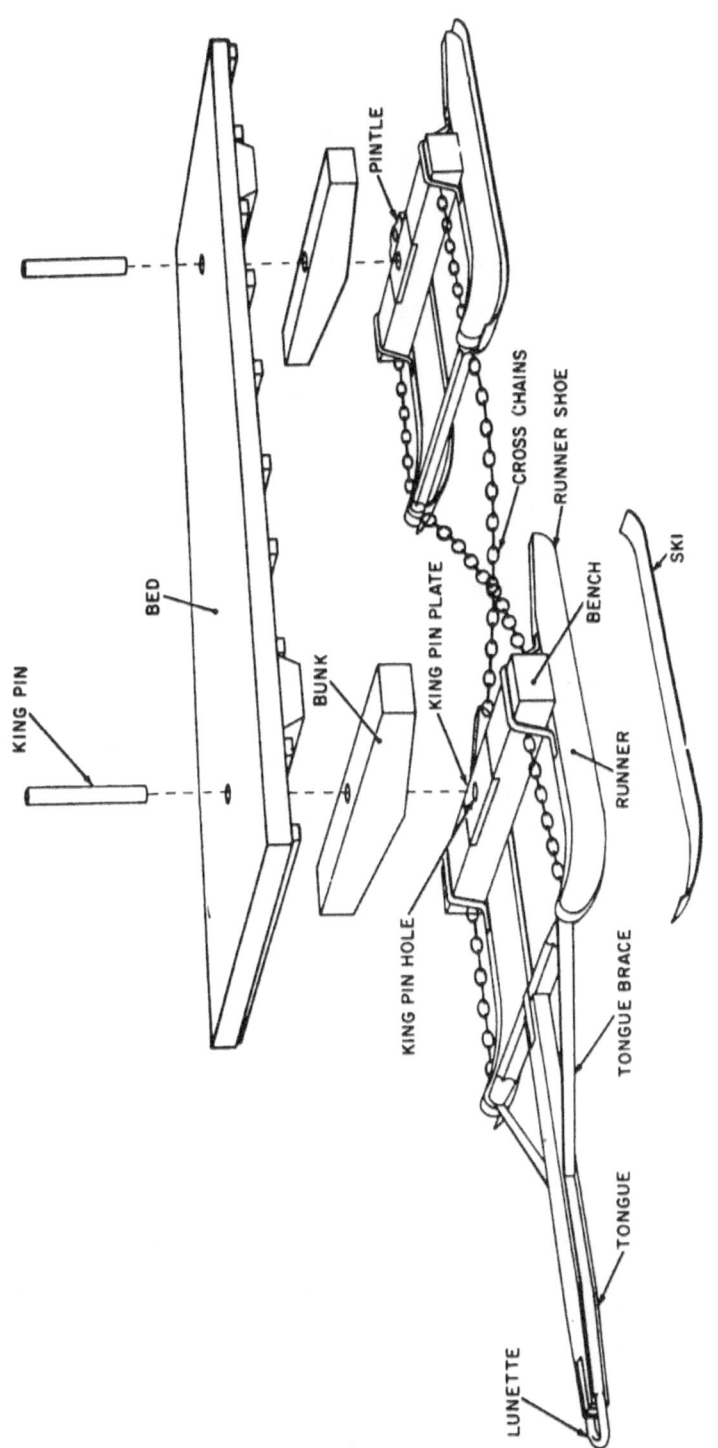

Figure 99. Representative drawing of heavy cargo sled.

Figure 100. The wanigan.

light sleds. The operating radius of sleds is restricted only by the terrain and capability of the prime mover. The heavy sled is best suited for use over flat or gently rolling terrain and in areas where rivers and lakes are frozen to sufficient depths to permit use as "highways." In some cases specially constructed "iced roads" are required to operate motorized sled trains with heavy sleds.

d. A wanigan is a sled with a housing structure built on the bed of the sled, or on a separate structure anchored securely to the bed. Wanigans presently in use resemble a railroad caboose and are of varying lengths, widths, and heights (fig. 100). Normally they are used for command posts, communication and radio shelters, kitchens, electrical generator shelters, sleeping quarters, first aid stations, and workshops. Wanigans may also be used for storage of supplies and equipment which should not be stored in the open.

Section VIII. AIRCRAFT

148. Aircraft

The lack of ground communication routes in the northern latitudes causes an extensive use of air transportation. Both fixed-wing and rotary-wing type aircraft are used. Troops and supplies may be transported from one existing or improvised airfield or airstrip to another. In some situations both supply and evacuation by air may be the only feasible method, especially under summer conditions. Bad weather may limit air operations for short periods of time.

a. Fixed-Wing. The vast stretches of the northern regions can be reconnoitered with a minimum time and effort by liaison fixed-wing aircraft. The ability of the ski-equipped aircraft to land on frozen lakes, streams, and in open fields in winter affords advantages and opportunities to supplement the ground reconnaissance. In addition to reconnaissance, fixed-wing aircraft are used successfully to supplement the overland movement of troops and supplies, evacuation, and many other purposes.

b. Rotary-Wing. The dominant characteristics of this type craft, such as vertical ascent and descent and requirement for short landing areas, make it valuable for reconnaissance, evacuation, troop movements, command control, resupply, and many other types of missions.

149. Airfields and Airstrips

There are many potential landing sites in the North. During winter, aircraft can land and take off on the numerous frozen lakes and rivers. Runways constructed by grading and compacting snow are used extensively. Aircraft may use skis instead of wheel landing gear. During summer, float-equipped airplanes and amphibious aircraft are used on the waterways. As a rule of thumb, the strip for liaison type aircraft (L–19 and L–20) should be a minimum of 90 feet wide and 1000 feet long. When the aircraft is ski-equipped, it requires longer space than when wheel-equipped. Both ends of the strip should be clear of obstacles. The landing strip at sea level for rotary-wing aircraft must be large enough to provide adequate clearance for the rotor blades and provide landing and takeoff paths reasonably clear of obstacles.

CHAPTER 6
COMBAT TECHNIQUES

Section I. THE INDIVIDUAL AND NORTHERN WARFARE

150. Problems of Northern Warfare

Two opponents face the soldier in northern warfare—the enemy, who must be defeated, and nature, which must be made an ally. We fight the enemy, but we must accept nature as it is, making nature fight with and for us. Proper clothing and equipment will help overcome the hazards of nature. Training teaches the individual how to use natural conditions for movement, concealment, and protection, as well as how to operate efficiently when the weather is good or bad, and in all types of terrain. The trained soldier moves, fights, lives, and works easily and confidently because he knows his job.

151. Nature of Northern Warfare

a. During winter the wide, empty spaces of the northern regions permit unrestricted maneuver and movement for troops sufficiently equipped and trained to operate in these circumstances. Dispersion is simplified; hostile artillery and mortar fire can be evaded or avoided. A mobile force can gain surprise and strike deep in the flank and rear areas of the enemy, disrupting his lines of communications and finally destroying him.

b. The principles of war remain unchanged. Tactics used in the northern latitudes are the same as anywhere else in the world. Northern warfare, however, consists of a great number of techniques to be used in summer and, especially, in winter operations. For the purpose of carrying out their mission, all individuals and units concerned must be indoctrinated and thoroughly trained in these techniques.

c. There is always opportunity for each soldier as an individual to display his initiative. Initiative is shown not only in combat, but also in the small things which can be done to make life more comfortable and more interesting in the North.

d. In the isolated areas of the North it is most essential that a system of teams be developed. Pair men together as "buddies" and insure a higher standard of efficiency, safety, and morale. If it can be avoided, never send one man alone on a mission—at all times try to keep "buddies" together.

Section II. INDIVIDUAL WEAPONS AND INSTRUMENTS

152. Effects of Northern Conditions on Weapons and Instruments

The year-round necessity for supervised care, cleaning, and maintenance cannot be overstressed. Effects of cold weather on various types of weapons are covered in detail in appendix V.

153. Care, Cleaning, and Maintenance

a. Weapons will function under extreme conditions, provided they are properly maintained. Normal lubricants thicken in cold weather and stoppages or sluggish action of firearms will result. DURING THE WINTER, WEAPONS MUST BE STRIPPED COMPLETELY AND CLEANED WITH A DRY CLEANING SOLVENT TO REMOVE ALL LUBRICANTS AND RUST PREVENTION COMPOUND. The prescribed application of special northern oils should then be made. These lubricants will provide proper lubrication during the winter and help minimize the freezing of snow and ice on and in weapons.

b. The individual soldier must insure that snow and ice does not get into the working parts, sights, or barrel of the weapon. Even a small amount of ice or snow may cause malfunction of the weapon. A rifle stand should be constructed (fig. 39). Muzzle and breech covers should be used. Before firing a weapon it must be examined carefully, especially the barrel, which may be blocked with ice or snow and will burst when fired. Snow on the outside, if not removed, may drop into the breech and later form ice, causing malfunctioning of weapon.

c. Condensation forms on weapons or instruments when they are taken from the extreme cold into any type of heated shelter. This condensation is often referred to as "sweating." When the weapons or instruments are subsequently taken into the cold, the film of condensation will freeze. The ice so formed may seriously affect the operation of any weapon or instrument. For this reason they must be left outdoors or stored in unheated shelters. If there is not too great a difference in temperature, weapons may be taken inside but placed near or at the floor level where the temperature will be lower.

d. When weapons are taken into a heated shelter, "sweating" may continue for as long as 1 hour. Therefore, when time is available, men should wait 1 hour and then remove all condensation and clean the weapon.

e. During the breakup, freezeup, and summer conditions, the danger of rust and corrosion is at its greatest. In the winter the lack of moisture in the air decreases this danger, but the problem of ice and snow will necessitate frequent checking and cleaning of weapons.

f. Should parts of a weapon become frozen, warm them slightly and move them gradually until unfrozen. If the weapon cannot be warmed, all visible ice and snow should be removed and parts moved gradually until action is restored. Ice in the barrel can be removed with warm (standard issue) gun oil if slow warming is not possible.

g. When firing, do not let the hot parts of the weapon come in contact with the snow. The snow will melt and, on cooling, form ice. When changing barrels, do not lay them on the snow; *rapid cooling may warp them.*

154. Ammunition

Extreme cold does not materially affect the accuracy of weapons nor the performance of small arms ammunition. Ammunition should be kept at the same temperature as the weapon. It should be carried in the bandoleers and the additional ammunition placed in the pockets of the outer parka and in the rucksack. Ammunition, clips, and magazines must be cleaned of all oil and preservative and must be checked frequently; all ice, snow, and condensation should be removed. Cartridge containers, magazines, and ammunition drums must be kept closed in order to prevent the formation of rust or ice.

a. Ammunition should be stored in its original container, raised off the ground, and covered with a tarpaulin. Ammunition so stored should be suitably marked in order to locate and identify it in the event it becomes covered with snow.

b. Resupply of ammunition may be restricted. All personnel must be made aware of the necessity for ammunition economy and fire discipline. Loaded clips, magazines, or single rounds dropped into the snow or soft summer ground are quickly lost; therefore, careful handling of ammunition is essential.

155. Care and Maintenance of Special Items

a. The liquid in the lensatic compass in extreme cold will become thickened. The heavy liquid slows the action of the compass and may make it inaccurate. This type of compass should be carried near the body in the inner clothing in order to keep the liquid warm and thin. The dry-type compass is not affected by cold weather.

b. Binoculars and other liquid-free optical instruments are not affected by cold weather. However, condensation does form when these instruments are taken from cold air into warm air. Therefore, as for weapons, instruments should be left outside.

c. Extreme cold will lower the efficiency of all batteries and eventually they may freeze. Batteries must be kept from freezing and, if possible, men should carry radio and flashlight batteries close to the body in order that full efficiency will be available when needed.

Section III. FIRE AND MOVEMENT

156. Blowing Snow and Fog

a. These restrictions will affect both friendly and enemy forces. Full advantage must be taken of them in order to effect concealment, surprise, and eventual success.

(1) Defense positions should be located on high ground, thus forcing the enemy to attack uphill in deep snow. Each weapon must be assigned a field of fire and emplaced on an improvised platform which will insure fire being brought to bear at man-height level on the likely enemy approaches. Thus, during fog, storm, or darkness, effective unobserved fire can be brought to bear.

(2) In areas of fog, if possible, outpost and observation post positions should be located where warmer air or wind eliminates fog or at least makes it less dense.

b. By proper reconnaissance and the use of trail markers it may be possible for an attacking force under cover of fog or blowing snow to approach very close to the enemy before the final assault. During blizzards or blowing snow the attacker should, if possible, attack downwind or at a slight angle to it in order that he will force the enemy to face into the full force of the storm.

c. Ice or vapor fogs are very common in extreme low temperatures. Such fogs are the result of many causes such as vehicle exhausts, cooking, breathing, and weapons firing. Fogs of this nature hang overhead and could be clear markers of a position. They will also limit visibility. The observed fire of automatic weapons and heavier caliber guns is handicapped considerably by the fog, smoke, and whirling snow caused by muzzle blast. Placing observers away from the weapons positions may be necessary to control the fire. Placing tarpaulins under the guns, or packing or icing the snow, will assist in reducing the effect of muzzle blast. Pauses in firing or change of position may be necessary in order to obtain better fire effect.

157. Fire Positions in Summer and Winter

a. Digging firing positions in soft or hard snow is relatively easy and quick. In a static position every effort must be made to improve the position and, if time permits, to dig it into the frozen ground. The use of explosives to dig emplacements and fires to thaw the ground will help. A position in the snow is only temporary and cannot withstand artillery and continuous small arms fire. Icing of the position or use of tree trunks and branches will afford added protection (fig. 101). Sandbags filled with snow may be used quite effectively for this purpose.

b. In the Far North during the summer, the almost completely frozen ground (permafrost), or water where it has thawed, will prevent the digging-in of positions. Therefore the positions will have to be built up above ground level.

c. The digging of positions in snow and the types constructed are, in general, similar to those discussed in FM 5–15. Foxholes, trenches, and other types are used.

d. Every effort must be made toward improvement of positions; snow blocks, ice blocks, sandbags, logs, and branches can be used to strengthen them. In addition, water may be poured onto the snow to form ice. In static positions, when time allows, water mixed with dirt, sand, or gravel can be poured into wooden forms. This is called "icecrete." The icecrete must be well tamped as it is poured to make it compact. Usually there is no necessity for removing the forms unless the wood is required for other purposes. Icecrete is darker than ice and will absorb more heat from the rays of the sun, causing melting. Icecrete construction must therefore be covered with snow, both to overcome its melting and to camouflage its contrasting color. Icecrete is much stronger than ice, provides considerable protection from small arms fire and shell fragments, and is a useful material for preparation of defensive positions. Icecrete, however, is brittle, and sustained fire reduces its protectiveness, thus requiring frequent repairs.

e. The action of winds and tides during winter rips the sea ice surface and then forces the ice into high piles extending in lines for miles. These ice barriers afford excellent firing positions

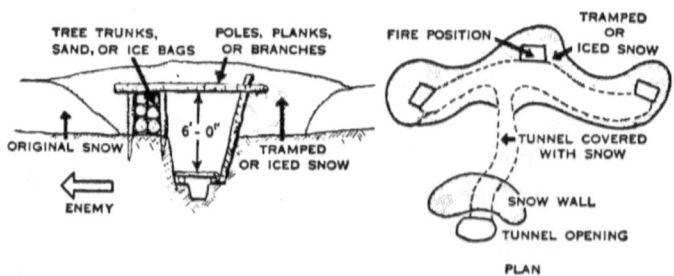

Figure 101. Example of position constructed in snow and extended into frozen ground.

and protection because of their thickness and the fact that they command the usually flat expanses between ridges. Ice blocks can be cut from numerous sources and used to strengthen a position. The ice should be covered with packed snow which will help camouflage and assist in eliminating the possibility of ricochets, shell fragments, and lethal ice splinters.

f. In a woods the thickest and strongest tree provides the best protection for the individual. In order to use the added protection afforded by the trees, perimeter positions should not be on the edge, but should be slightly deeper in the woods, depending on its density and consistent with the required fields of fire (fig. 102). A tree 20 inches in diameter will provide protection from small arms fire. If the tree selected is smaller, packed snow, dirt, branches, or deadfalls may be used to increase protection.

g. The improvement of fields of fire in woods is most important. The lower branches of trees, up to 6 feet high, which restrict fields of fire must be removed. Underbrush and perhaps even a few trees will have to be cut; however, do not strip the area. In the first phase of improvement, criss-crossing snow tunnels under the trees is carried out. Then, if time allows, those fields are extended wider and deeper. In the final phase, obstacles and traps are constructed and mines laid in these areas (fig. 102).

158. Use of Ski Poles and Sleds in Firing

a. When firing in snow, it is necessary that a firm support be used, as snow will compact. On hard packed snow the weapon may slide. Therefore, any item available in the area or in the men's possession should be used to insure a solid base; e.g., branches, skis, or sleds.

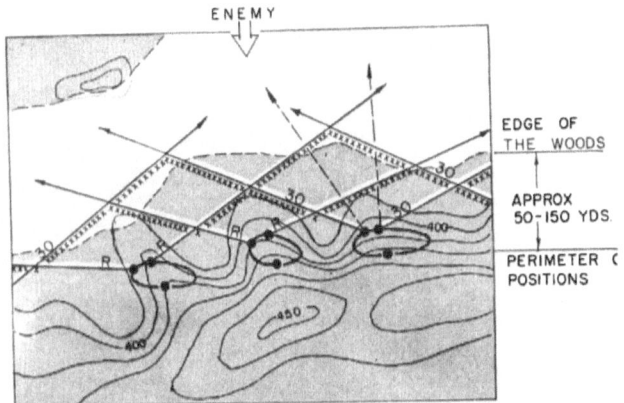

Figure 102. Perimeter positions at the edges of the woods, and clearing of fields of fire.

b. Skis and ski poles can be used in a variety of ways to form weapons rests while firing on the move. Figure 103 illustrates the standing position. Use this position only in hasty situations, as when surprised by enemy fire.

c. Ski poles may used as an elbow rest or as weapon support when firing from a kneeling position in shallow crusted snow (fig. 104). For firing in deep, soft snow the position of the poles can be reversed for added stability.

d. When firing from the prone position, the skis or ski poles may be used as an elbow rest or as supports for the weapon (figs. 105–107).

Figure 103. Standing position.

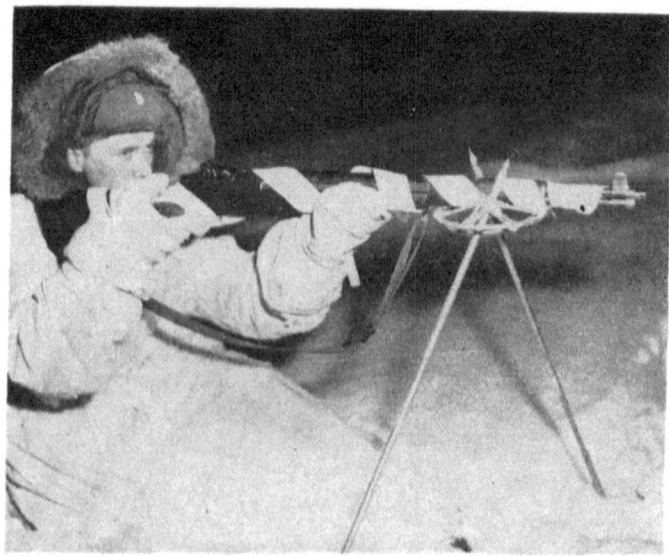

Figure 104. Kneeling position.

Figure 105. Prone position. Skis used as weapon support.

Figure 106. Prone position. Ski poles used as elbow rest.

e. Automatic weapons may be fired from the prone position using a snowshoe or ski pole basket as a rest for the bipod (fig. 108). A fairly wide strip of canvas may be permanently attached to each leg of the bipod. On opening the bipod, the canvas will stretch out between the legs over the snow and stop the legs from sinking.

f. To prevent weapons from sinking in deep snow, machine guns may be fired from sleds in case of emergency (fig. 109). The weapons can be mounted either with regular or improvised mountings. However, it is essential that weapons be placed in a dug-in position as soon as possible.

159. Strength of Snow, Ice, and Frozen Ground for Cover

a. General. The soft, spongy ground of the North in the summer, and the snow surface in the winter, have a smothering effect on fire from all types of weapons. Hard frozen, bare ground or ice, when not covered with snow, greatly increases the number of richochets and fragmentation effects. The resistance or protection offered by snow, ice, or frozen ground against enemy fire is variable.

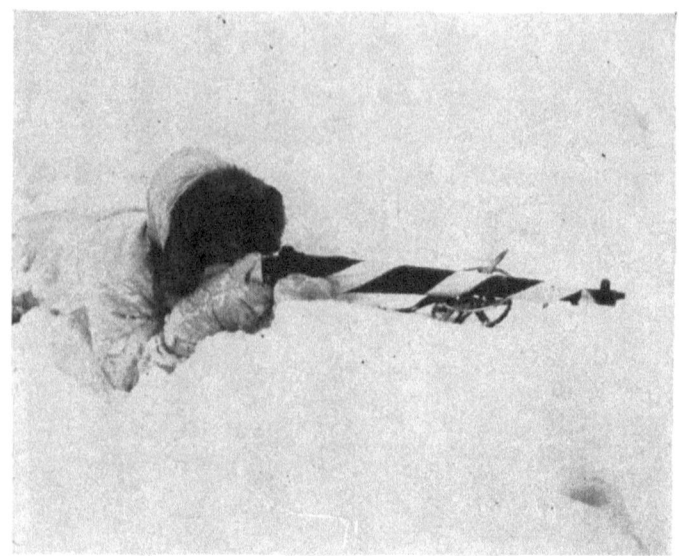

Figure 107. Prone position. Ski poles used as weapons support.

Figure 108. Use of snowshoe or ski pole basket as weapon support.

Figure 108—Continued.

Figure 109. Machine gun firing from a sled.

b. Penetration. A rifle bullet rapidly loses its killing power depending on the density of the snow. Snow packed in layers tends to deflect the bullet at each new layer. Loose snow spread over a defense position will help smother richochets. The minimum thickness for protection from rifle bullets and shell fragments is shown in the following table:

Penetration Table

Snow wall material *	Minimum thickness in feet
Newly fallen snow	13
Firmly frozen snow	8 to 10
Packed snow	6½
Frozen snow-water mixture	4 to 5
Ice	3¼
Icecrete	1

* These materials will disintegrate under sustained fire.

160. Effect of Snow, Ice, Frozen Ground, and Muskeg on Shells and Grenades

a. Loose snow greatly reduces the explosive and fragmentation effects of shells. The depth, type of snow, and ammunition are naturally the main considerations. The use of a delayed action fuze will generally cause the shell to penetrate the snow blanket and explode underneath, smothering and reducing the effect of the fragmentation. Three feet of snow will provide some protection against most light artillery fire. A superquick fuze setting will increase the effect of artillery fire, while air bursts will inflict still more casualties on surface targets.

b. In the summer the many areas of muskeg and water will also limit the effects of artillery fire. On ice or frozen ground, and during periods of freezeup, the effect will be greatly increased as the result of flying ice splinters and frozen clods of ground. In these seasons and areas, covered positions must be increased in strength. Overhead protection must be sought whenever possible.

c. Hand grenades often sink into the snow and muskeg before exploding and are consequently restricted in effect. The use of heavy mittens will decrease accuracy and throwing range.

161. Crew-Served Weapon Positions

a. Detailed information and guidance for construction of emplacements and shelters is contained in FM 5-15. The dimensions are applicable for both winter and summer. The gun emplacements for MG's and 3.5-inch RL's are square-type positions. The gun platform can be made from packed snow and is about waist high. Open space must be left behind the gun to allow for back blast of the rocket launcher.

b. Mortar positions in snow are normally round shaped (fig. 110). Due to frozen ground a mat made from tree branches or sandbags filled with snow is placed under the baseplate in order to prevent breaking the baseplate when firing.

c. Bunker-type positions will give better protection for the gun crew against enemy fire and weather than will open positions (figs. 111–113). A hasty bunker-type position is normally built as follows:

    (1) A square shaped hole is dug in the snow, the dimension depending on the purpose of the bunker position.

    (2) A heavy log or a tree trunk is placed lengthwise on each side of the snow hole. They are supported by four heavy, forked poles.

    (3) A layer of logs is placed crosswise in the top of the two support logs.

    (4) A layer of boughs is placed on the first layer of logs in order to prevent melting snow from dripping into the bunker.

    (5) Two or three more layers of logs are placed on the top of the boughs.

    (6) Finally, the roof is covered by smoothing and packing the snow in order to eliminate any sharp features that may produce shadows.

    (7) A small embrasure reinforced with sandbags and snow is left open, in the direction of the field of fire.

Figure 110. Mortar position in snow.

(8) The rear entrance is covered with a white tarpaulin or a white camouflage suit.

d. Tents are often used in temporary defense positions to shelter the men. They must be close to the combat positions and should be in defilade. The tents must be dug into the deep snow, or even into the ground, in order to protect the men against enemy fire. The tent ropes must be well anchored by using deadman anchors or upright poles placed deep in packed snow. Immediately outside the tent, defense positions must be dug for use in case of sudden alert (fig. 114).

e. When near the surface the covering snow is easy to dig with individual intrenching tools; the difficulties will start when frozen

Figure 111. Bunker-type position (from inside).

Figure 112. Bunker-type position (from the front; partially camouflaged).

Figure 113. Bunker-type position (from the front; camouflaged).

Figure 114. Tent dug into snow with individual firing positions.

ground is reached. Several small holes should be dug in the ground and attempts made to break the frozen ground between them. The men should temporarily exchange the different types of individual intrenching tools in order to make the digging faster. During darkness, or in areas not under the enemy's direct observation, heavy tools such as picks, crowbars, and shovels are used so that positions can be completed rapidly.

f. Using explosives is the easiest and fastest way to break the frozen ground. However, the use of demolitions will be restricted when under enemy observation. TNT is the best type of explosive for this purpose because it is less sensitive to the effects of cold weather. Dig a hole in the ground big enough to place the TNT charge, place the cap and time fuze into the charge, cover it well, and detonate. By using electric caps and electric circuits, several holes can be blasted simultaneously. For an individual soldier's foxhole, 10 lbs of TNT will normally be sufficient. Another proven formula is to put two 1-lb blocks of TNT for every foot of penetration in frozen ground. A very efficient method of making bore holes is the use of shaped charges as described in TM 5-560.

g. Some improvised means as listed below may be used to break the frozen ground when no others are available:

    (1) In rear areas frozen ground can be thawed by starting a campfire in the place where it is desired to dig.

    (2) Two or three hand grenades tied together can be used to blast a hole in the frozen ground.

    (3) Existing craters caused by enemy or friendly artillery fire can be utilized.

    (4) Often the tops of ridges or hilltops will be rocky and with very little snow on the ground. If the time and situation allow, the snow situation can be improved by erecting snow fences in the place planned for defense positions. Within a few days the snow fences will collect drifting snow in banklike forms in which it is easy to dig positions.

Section IV. FIGHTING TECHNIQUES

162. Formations

Squad and platoon formations for tactical movements remain basically the same as for temperate regions; however, terrain and deep snow cover will necessitate some modifications. In deep snow, when speed is of the essence, a column formation may be preferable to a line formation because it will require fewer ski trails. Old, well-settled snow will normally provide good flotation and will facilitate skiing for the individuals. Since the trailbreaking requirement is reduced and may under favorable circumstances be nonexistent, line formations may be used without loss of speed. Downhill movement, even in deep snow, may also indicate the use of line formations when it would not be considered feasible on level terrain under the same snow conditions.

163. Handling of Ski and Snowshoe Equipment and Individual Weapons

a. The purpose of using skis or snowshoes in combat is to expedite the movement of individuals over deep snow in the most rapid manner, thus exposing them to hostile fire for the shortest possible period of time. In order to obtain the maximum advantage of skis they should be used as far forward as possible, leaving them behind only when the enemy definitely can be reached more quickly and easily on foot. It is finally up to the small unit leader to decide at which phase in the attack this may be done. As a rule of thumb the skis are left in the assault position, as close combat on foot is more effective and easier to execute than if mounted on skis. Conversely, deep snow may force units to close into the objective on skis.

b. As friendly forces approach the effective range of enemy weapons, they move by fire and maneuver. The individuals proceed by short rushes on foot or on skis, whichever is most feasible. Rushing on foot, the skis are dragged by holding them together by the tips (poles through the toe straps) in one hand, with the weapon readily available for action in the other (fig. 115). Skis may also be tied to the belt with the emergency thong slipped through the holes at the ski tips.

c. To be able to quickly dismount from skis when hostile fire becomes effective, the ankle straps of the cross-country bindings should be opened. When advancing by short rushes on skis, only toe straps are used. Under favorable snow conditions, as well as in emergencies, the ski bindings are kept on when lying down and firing between rushes (fig. 116).

d. The individual weapon is carried across the back with the sling over either shoulder, the butt at the side or attached to the man-pack (if carried by the individuals), when contact with the enemy is not expected (fig. 117). The weapon is slung around the neck and in front of the body when contact with the enemy is

Figure 115. Dragging of skis in rush.

imminent, thus releasing both arms for rapid skiing (fig. 118). When contact with the enemy has been established, the weapon is carried in one hand and the ski poles in the other so the weapon is readily available for action (fig. 119).

e. Individuals using snowshoes may keep them on through all phases of attack. Under favorable snow conditions they may be left piled together in the assault position or carried by the individuals in the final assault, fastened to the individual's equipment where they will least bother him.

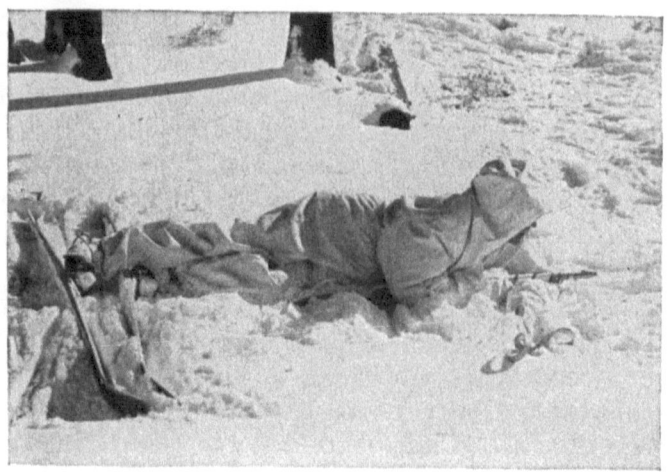

Figure 116. Lying down between rushes.

Figure 117. Weapon carried across back.

Figure 118. Weapon slung around neck.

Figure 119. Weapon carried in right hand and ski poles in left hand.

f. Under conditions where the depth of the snow is less than 1½ feet, skis may be left in the attack position if it becomes evident that launching an attack on foot can be executed in a more rapid and efficient manner than using skis.

g. As soon as the objective has been seized, the skis, ski poles or snowshoes may be recovered and brought forward. A two-man team can quickly make a ski bundle (fig. 120) and drag the skis of an entire squad at one time.

Figure 120. Ski bundle.

164. Additional Techniques

a. In deep, loose snow under hostile fire it may be more advantageous to advance in a crouching position by holding the skis with hands through the toe straps and taking full advantage of snowdrifts and bushes. A position such as illustrated in figure 121 should be adopted. Snowshoes may be used in the same manner.

b. Sliding forward on skis is another method of advancing, especially over firm snow (fig. 122). The rifle can be slung over the shoulder or laid on the skis directly in front of the individual. The latter is possible only when the snow is hard so that it cannot get into the rifle.

c. In deep snow, trenches may be dug in the snow leading in the direction of objective when it is too difficult to be reached by oversnow movement. Snow trenches are dug on a zigzag course (fig. 123) by throwing the snow out under cover of darkness or, in an emergency, the digging may be masked by smoke screens. The snow shoveled from the trench should be placed on the enemy side of the trench to allow the individuals to crawl along the trench without being observed by the enemy.

Figure 121. Advance in crouching position.

Figure 122. Sliding forward on skis.

Figure 123. Snow trenches.

Figure 124. Action of wind.

  d. Snowdrifts and vehicle tracks may be utilized when found in the battlefield. Snow fills in ditches and rolling ground and tends to flatten the terrain in general. The wind builds up snowdrifts and cornices and can change the contour of the ground a great deal. Snow-covered terrain must be continually studied and every feature utilized. On the downwind side of every obstacle, tree, house, and bush there is always a hollow which may provide an excellent observation point or firing position (fig. 124).

  e. The wind, particularly in open areas, may form long, wavy snowdrifts which are almost natural snow trenches. They may at times be used as an approach to the objective.

  f. Frozen streams or sunken riverbeds may be used as another means of advance (fig. 125); often they may represent a longer but safer route.

Figure 125. Sunken river bed.

Figure 126. Snowbanks beside plowed road.

g. An early fall frost will form a layer of ice on creeks or streams when the water level is high. Later, when the flowing water becomes lower and reaches its winter level, the top surface will again freeze so that there are two layers of ice. This is called shell ice and is not safe.

h. Certain swampy areas do not freeze solidly during the coldest periods of winter. They are often covered with snow, hiding the water underneath and making the swamps an obstacle. Only experience and the knowledge that they exist in the local area, will prevent accidents. Suspected areas should be avoided.

i. Snowbanks beside plowed roads and tracks often provide excellent cover in wintertime (fig. 126). These banks or drifts will remain far into the spring thaw period, especially in areas of heavy snowfall.

j. The tracks left by tanks and oversnow vehicles in snow may provide routes of advance (fig. 127). Continuous traffic packs the snow and may allow movement on foot without skis or snowshoes. In the advance, infantry may utilize tracks left by their advancing armor.

k. In static situations the ski equipment becomes vulnerable to small arms fire and shell fragments. When troops are expected to remain in the same position for an extended period of time, skis, poles, and snowshoes should be placed in a covered position.

Figure 127. Tracks left by tank.

Section V. CAMOUFLAGE AND CONCEALMENT

165. General Considerations

a. In winter the whiteness of the countryside emphasizes any item which may not blend in naturally with the surroundings. Furthermore, every movement by vehicles or dismounted troops leaves tracks in the snow. Before every movement, consideration must be given to how these tracks can be kept to a minimum. Nature may assist by covering tracks with newly fallen snow or by providing a storm in which all movement will be concealed. Camouflage and concealment from air observation is of the greatest concern.

b. In the northern landscape, backgrounds are not necessarily all white. Rocks, scrub bushes, and shadows make sharp contrast with the snow.

c. Snow-covered terrain in the wooded regions, when viewed from the air, reveals a surprising proportion of dark areas.

166. Vapor Clouds

Firing of weapons, vehicle exhausts, and breathing will, in extreme cold, cause local fogs or vapor clouds which can be seen by the enemy even though the weapon, vehicle, or soldier is well concealed. Smoke from fires hangs immediately above and will disclose the position if there is no wind to blow it away. Under certain conditions, if the position is on a high point, smoke may flow downward into depressions and may be used as a deceptive measure. It may be necessary to move weapons frequently, shut off vehicle motors, or leave vehicles in rear areas. Wood fires should not be allowed by day in open areas. Conversely, deception or concealment might be gained by deliberately creating vapor fogs or clouds.

167. Sounds

The still, cold air of the North carries sound much farther than in temperate climates. All sounds must be kept to a minimum. Noise caused by motors, men coughing, and skiers breaking through snow crust may warn the enemy of activity at extreme distances.

168. Visibility

The long hours of daylight in the North during the summer allow for longer periods of aerial reconnaissance and increase the possibility of detection. The short hours of daylight during the winter months materially decrease the time available for reconnaissance.

169. Tracks

a. Tracks made in a soft surface may become quite firm if the temperature drops during the night, and will remain indefinitely as indications of movement. In the summer, tracks across spongy masses become quite clear to the aerial observer. Special consideration must be given to the tracks in bivouacs and base camps. Number and size of trails must be kept to a minimum. All unnecessary "streets," turnaround loops, and parking areas must be avoided. Individuals may be forced to use only a certain trail by stringing single barbed wire on both sides of the trail. From the air, tracks, even through wooded areas, appear like a white scar. Coniferous branches can be laid in a staggered pattern on each side of the track as well as on it. Strict track discipline both during movement as well as in bivouacs and base camps must be maintained at all times.

b. Aerial photographs are closely examined and from them can be gathered a great deal of information. The depth of a track will show the amount and the direction of movement. Vehicle or sled tracks may indicate the type of vehicle and conclusions can be made as to the type of weapons. Every effort must be made to mislead the enemy. It may be advantageous to make more tracks or trails and show greater signs of strength. All marks made in the open are generally visible to the camera. Muzzle blasts, refuse, or light reflected from shiny surfaces may easily reveal positions. During the northern nights any light from a tent or fire may be seen for a considerable distance.

170. Camouflage Materials

a. White is the predominant color in winter and snow is the most important camouflage material. By intelligent use of camouflage clothing and equipment together with what nature makes available, effective individual and group camouflage can be achieved.

b. Improvised camouflage clothes can be made from sheeting, tape, whitewashed sacking, or painted canvas. White paper, when wet, can be applied and allowed to freeze on all kinds of surfaces. Snow thrown over the object helps to increase the camouflage effect.

c. White paint has many uses in winter camouflage. Weapons, vehicles, skis, and sleds can be effectively painted with white nonglossy paint.

d. On occasion, white smoke may be used to help the camouflage plan. The major problem is to make the installation blend in with the countryside.

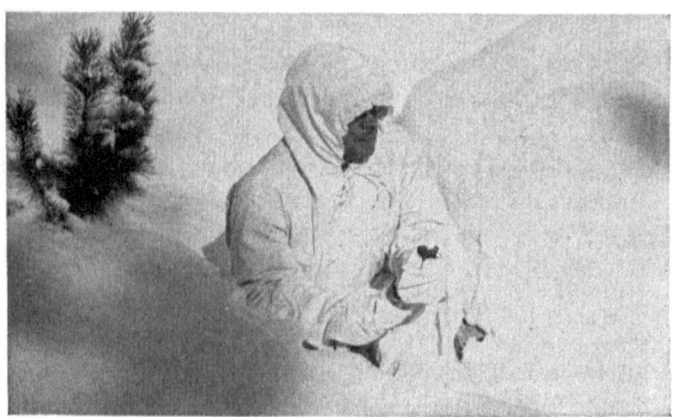

1 Observer.

2 Squad on mountainside.
Figure 128. Camouflage in open snow conditions.

171. Individual Camouflage and Concealment

a. During the summer the normal principles of using camouflage clothing will apply. However, as winter approaches, men must use partial white winter camouflage to match the changing conditions; men should be trained to avoid areas of local growth and dark outlines (fig. 128).

b. In fairly open forest areas during the winter, men wearing "whites" should avoid the dark background of trees. In the same manner, if wearing dark clothing, men should stay under trees and avoid the open.

1 Right

2 Wrong

Figure 129. Use of mixed clothing.

c. In mixed surroundings frequent changes of camouflage clothing become necessary. The use of mixed clothing is often the most preferable (fig. 129).

d. All equipment worn on the outside should be camouflaged. Contrasting equipment worn on the camouflage suit will increase the possibility of enemy detection. Loose items such as grenades or field glasses should be kept concealed inside the suit.

172. Camouflaging Equipment

Skis, rifles, and sleds may be painted white prior to issue. If they are unpainted, white camouflage paint or improvised local materials can be used. Sleds wil be issued with white covers for concealing the load. Finally, individual weapons can be camouflaged with strips of white garnish or white adhesive tape. The tape also provides protection for the hands when handling the weapon in extreme cold.

173. Camouflage and Concealment of Small Groups

a. In selecting a position, enemy ground and air observation must always be considered. A location which requires the least amount of modification is the most suitable, since there is less requirement for disturbing its "natural" appearance. The camouflaging of a position commences before occupation of the position. The most suitable covered approaches must be used and tracks, if not hidden, must be kept to a minimum. Where possible approaches should be made under cover of trees or bushes, behind snowdrifts or slopes, and in shaded areas. Poor camouflage at this point may make position camouflaging ineffective. If tracks cannot be concealed, then tracks should lead through the position to one or more dummy positions. On occupation of a position, disturb its appearance as little as possible. Snow or earth removed from the position should be thrown to the enemy side. If the position is of snow or ice construction, it must be rounded off in order to avoid reflection and marked shadows. Overhead tarpaulins or camouflage nets should be used to cover any extensive digging in snow or earth.

b. In placing the individual and the weapon it is most important that he is not silhouetted or contrasted with his background. Low positions that blend into the background is the secret.

c. If time allows, positions can be greatly improved by constructing an overhead cover of suitably camouflaged materials such as branches, nets, blankets, etc. (fig. 130).

d. The tent is probably one of the largest items to be camouflaged (fig. 131). It is the main protection against the elements. Al-

Figure 130. Covered foxhole in snow.

though large, by careful site selection using both artificial and natural camouflage material, it can be readily hidden. A decreased number of tents and stoves, due to tactical reasons, will automatically assist in keeping the bivouac area camouflaged. Occasionally, the camouflage of the tents in sparse vegetation, barren tundra, and especially under winter conditions becomes very difficult. Use white materials such as individual overwhites or snow blocks to protect the dark material from observation. In emergencies the white inside liner may be removed and placed on the top of the tent. Frequently all fires in the stoves as well as the open fires must be extinguished and the warming factor sacrificed for camouflage and safety reasons.

174. Camouflage of Vehicles

a. In the summer the camouflaging of vehicles is effected by normal procedures. In winter all vehicles should be painted white to fit the predominantly white terrain. In forested areas it is relatively easy to darken a white vehicle with issued or improvised camouflage material. In areas with definite contrasts, for example

Figure 131. Camouflage tent in snow.

in the wooded areas, or during breakup and freeze-up periods, a mottled effect should be used.

b. In addition to the vehicle painting, each vehicle should be equipped with an all seasonal camouflage net (fig. 132) to be used when required. Concealment will be more effective if vehicles are parked close to dark features or in shaded areas (fig. 133). Always try to break the silhouette of the vehicle and avoid vehicle shadows. Try to make it appear flat when observed from the ground or air.

A. White vehicle without cover

B. White cambric cover conceals same vehicle

Figure 132. Vehicle camouflage.

Figure 133. Comparison of visibility of vehicles and sleds exposed to the sun's rays and vehicles parked in shadows.

  c. In wooded areas vehicle lean-tos can be built to conceal vehicles. In a static situation a snow shelter can be constructed to provide cover and concealment.

  d. In extreme cold consideration must be given to the exhaust from vehicles since it will form ice fog and provide the enemy with additional means of detection.

175. Deception

  a. More opportunities for unit or individual deception exist in the North during winter than possibly in any other area. However, deception measures are not sufficiently effective to lessen the requirement for good concealment. Unless unit and individual camouflage is effective, the value of any deception plan will be greatly reduced. Deception must be based on well-coordinated

Figure 134. Deception area, showing trails and tracks in forest and dummies in open areas.

plans which must be logical and not too obvious. Dummy positions must be positioned to follow the tactical plan, but far enough removed from actual positions so that fire directed at the dummy position will not endanger the real position (fig. 134).

 b. A few skiers or oversnow vehicles can create a network of trails or tracks to mislead the enemy as to direction, strength, location, and intentions.

 c. Regular pneumatic deception devices are inoperable and should not be used in temperatures below zero degrees. Improvised devices, however, can be made from snow, branches, canvas, and any other available material. Dummy weapons, positions, tents, and vehicles of all kinds can be constructed (fig. 135). They must not appear obvious but should appear camouflaged and only "discovered" as a result of a camouflage violation. A dummy bivouac area must appear to be occupied. Small gasoline or oil flames may be used to simulate stoves or idling engines. In a bivouac area (fig. 136) the place must appear to look occupied; a fire or smoke could easily be used to produce this effect.

Figure 135. Pine branches used to provide dark texture to simulate a weapon position.

Figure 136. Dummy bivouacs and locations.

Figure 137. Snow dummy of an oversnow vehicle.

d. In accordance with the deception plan, a landing strip can be easily and quickly established, utilizing one oversnow vehicle and a few men.

e. All types of dummies can be built from snow blocks or piled up snow (fig. 137). Snow can be quickly shaped with a shovel, and detail can be added by carefully placing spruce branches, dirt, ashes, logs, and any other suitable material. The necessary vehicle tracks must be made.

f. Another method of constructing dummies is to erect a frame of boughs and fill it in with branches. The uprights can be frozen into the snow or ground by pouring water around the base, and the cross members may be lashed together. Snow should then be sprinkled over the frames and partial camouflage carried out. A net may also be used.

g. Detailed information pertaining to camouflage in the northern latitudes is covered in TM 5-560.

Section VI. MINES AND OBSTACLES

176. Use of Antitank Mines

a. Antitank mines must be placed on a *solid base,* otherwise when pressure is applied they will sink into the soft ground or snow and lose much of their effectiveness. In shallow snow a hole may be dug and the mine placed on the frozen ground. In deep snow they must be supported. Additional charges will help overcome the smothering effect of deep snow. The snow may be tramped down or frozen, or the mine may be placed on a plank or something similar to provide the required firm support (fig. 138). In all cases they must be covered with snow or dirt, but not buried too deeply; otherwise the top layer may accept the weight and not detonate the mine. A piece of cardboard over the mine will protect it from moisture which may freeze and hinder the working parts.

b. All mines should be painted white, as the wind may blow the covering snow away. The mine or mine field should be carefully recorded and marked, as new snow may change the complete appearance of the terrain.

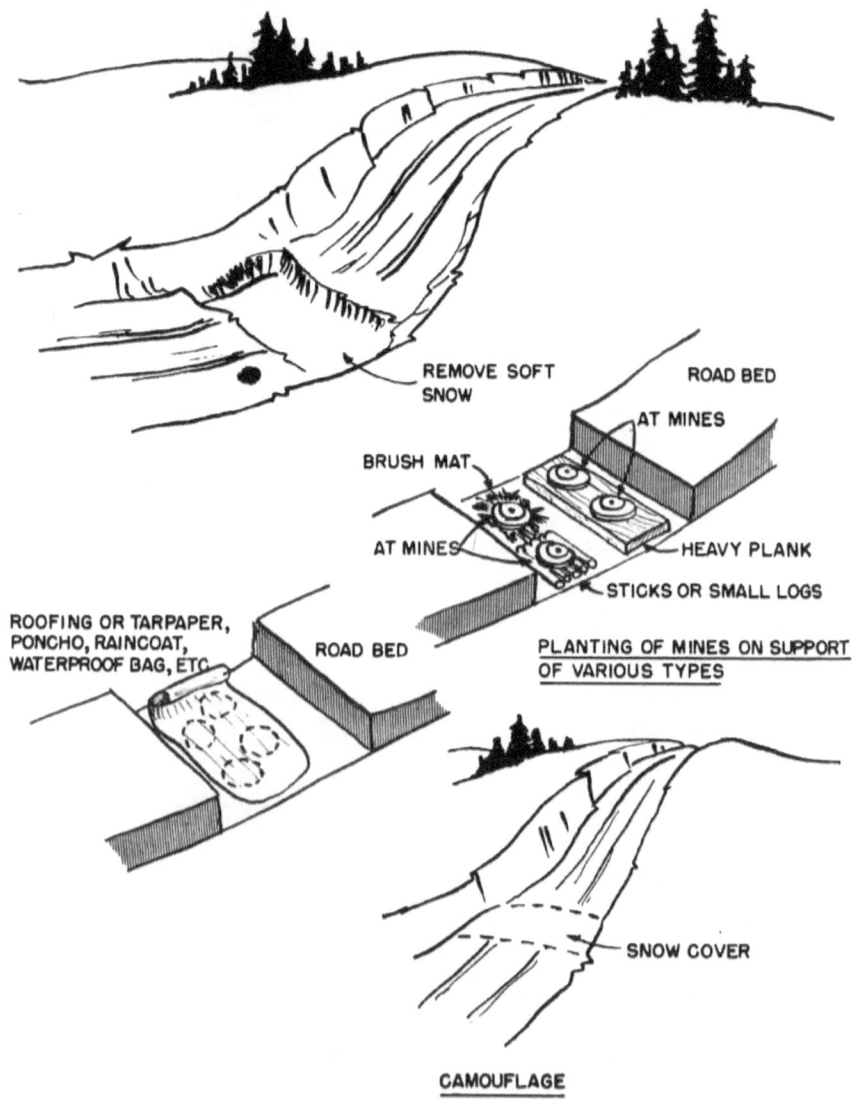

Figure 138. Laying mines in deep snow.

177. Antipersonnel Mines

a. If using pressure-type igniters, a solid support is necessary. When using pull-action igniters, white trip wires are necessary in winter. The trip wires may be placed in the snow or just above it. In forested areas the mines are placed on tree trunks and the wire stretched across to another tree. They, too, must be protected against freezing moisture which may make them ineffective.

b. Mines are used on ski or snowshoe trails (fig. 139). Because the weight of the individual is distributed along the length of the ski or snowshoe, these mines must be more sensitive to pressure. This type of mine can be easily improvised. Place an explosive into a moisture-proof container and use an internal pressure device together with a pressure fuze. Place the mine about 1 inch under the surface.

178. Use of Demolitions in Ice

a. General. Thousands of lakes, rivers, and swamps found in the northern regions, in summer provide formidable obstacles to armor and personnel. In winter, however, when frozen to sufficient depth, they provide excellent avenues of approach. They also lengthen the front line of a given sector, requiring more troops and weapons to defend it than in summer. Necessary action must be taken to deny these natural routes to the enemy under winter conditions.

b. Laying Mine Fields in Ice.

    (1) In order to create water obstacles during winter conditions, explosives are used to blow gaps in lake and river ice impassable to enemy personnel and armor. To install the demolition in ice (fig. 140), holes are sunk ten feet

Figure 139. Placing mine on ski track.

apart in staggered rows by uses of axes, chisels, ice augers (fig. 141), steam point drilling equipment, or shaped charges. The shaped charges will not make a hole large enough to pass the charge through but must have the hole widened by other means. Charges are suspended in the water below the ice by means of cords tied to sticks bridging the tops of the holes. The charges should be TNT, Tetrytol, or C–4. The C–4 should be protected against erosion by water currents. Demolitions laid early in winter must be placed deep enough so that they will not be encased in the ice as it grows thicker.

(2) The normal thickness of fresh water ice is approximately four feet or less. In extremely cold areas five feet of ice is not uncommon. At the time the mine field is established, it is difficult to determine how thick the ice will be at the time the mine field is detonated. As a rule of thumb, if the ice is expected to be four feet thick the charges should be approximately 10 lbs. In the event the depth of the ice is expected to exceed four feet, an addition of 2.5 lbs. per additional foot of thickness should be emplaced. Electrical firing devices are attached to three charges in each underwater demolition, one in each end charge and one in the middle charge. The rest of the charges may be primed with concussion detonators or electrically primed. The large number of charges does limit the use of electrical means of firing. An ice demolition may consist of several blocks of charges echeloned in width and depth and has at least two rows of mines, each row alternating with the one before it. Blowing a demolition such as this creates an obstacle for enemy armor and vehicles for approximately 24 hours at —24° F. (FM 5–25 and TM 5–560).

(3) Great care must be exercised when handling electrical firing devices under winter conditions. Because of improper grounding of an individual, due to snow and ice covering the ground, the static electricity that builds up might possibly detonate the device. Individuals must insure they are properly grounded prior to handling any type of electrical firing devices.

c. *Advantages.*
(1) Long sectors of the front line may be cut off at a critical moment from enemy infantry and armor.
(2) Number of personnel and AT weapons needed to defend a given sector is reduced.

(3) Friendly troops may advance or withdraw at any place over the charges without being restricted to the cleared lanes.

(4) Charges laid under thick ice are difficult, and often impossible, to detect by use of mine detectors.

(5) When the holes over the charges have refrozen, the field is very difficult for the enemy to breach.

(6) The charges are not affected by weather or snow conditions.

(7) After a snowfall, detection of the demolitions by the enemy is extremely difficult.

 d. *Disadvantages.*
(1) Emplacing the explosives requires considerable time even when ice cutting equipment is available.

(2) The charges can be set off when hit by artillery fire.

(3) The gaps blown in the ice tend to freeze over rapidly in low temperatures.

 e. *Tactical Use.* Ice demolitions are used for protection from frontal or flanking attacks. Normally, one or more sets of charges are laid close to the friendly shore and others farther out in the direction of the enemy (fig. 142). If desired, the enemy may be allowed to advance past the first set of charges and then both detonated at the same time. The enemy thus will be marooned on an ice floe, unable to continue to advance or retreat, and can be destroyed. The same trapping method may be used against enemy armor, or the charges may be detonated directly under the advancing tanks. Ice demolitions must be kept under observation and secured by friendly fire.

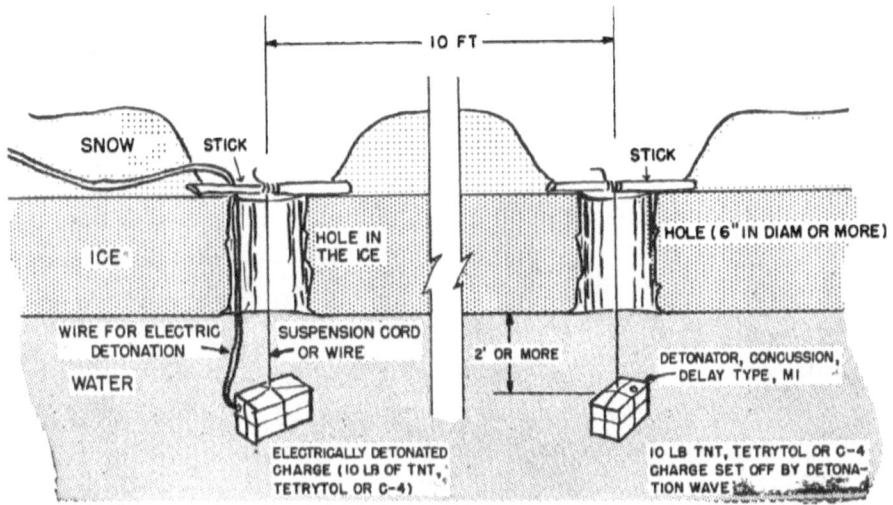

Figure 140. Method of placing charges in ice.

apart in staggered rows by uses of axes, chisels, ice augers (fig. 141), steam point drilling equipment, or shaped charges. The shaped charges will not make a hole large enough to pass the charge through but must have the hole widened by other means. Charges are suspended in the water below the ice by means of cords tied to sticks bridging the tops of the holes. The charges should be TNT, Tetrytol, or C-4. The C-4 should be protected against erosion by water currents. Demolitions laid early in winter must be placed deep enough so that they will not be encased in the ice as it grows thicker.

(2) The normal thickness of fresh water ice is approximately four feet or less. In extremely cold areas five feet of ice is not uncommon. At the time the mine field is established, it is difficult to determine how thick the ice will be at the time the mine field is detonated. As a rule of thumb, if the ice is expected to be four feet thick the charges should be approximately 10 lbs. In the event the depth of the ice is expected to exceed four feet, an addition of 2.5 lbs. per additional foot of thickness should be emplaced. Electrical firing devices are attached to three charges in each underwater demolition, one in each end charge and one in the middle charge. The rest of the charges may be primed with concussion detonators or electrically primed. The large number of charges does limit the use of electrical means of firing. An ice demolition may consist of several blocks of charges echeloned in width and depth and has at least two rows of mines, each row alternating with the one before it. Blowing a demolition such as this creates an obstacle for enemy armor and vehicles for approximately 24 hours at —24° F. (FM 5-25 and TM 5-560).

(3) Great care must be exercised when handling electrical firing devices under winter conditions. Because of improper grounding of an individual, due to snow and ice covering the ground, the static electricity that builds up might possibly detonate the device. Individuals must insure they are properly grounded prior to handling any type of electrical firing devices.

c. *Advantages.*

(1) Long sectors of the front line may be cut off at a critical moment from enemy infantry and armor.

(2) Number of personnel and AT weapons needed to defend a given sector is reduced.

(3) Friendly troops may advance or withdraw at any place over the charges without being restricted to the cleared lanes.

(4) Charges laid under thick ice are difficult, and often impossible, to detect by use of mine detectors.

(5) When the holes over the charges have refrozen, the field is very difficult for the enemy to breach.

(6) The charges are not affected by weather or snow conditions.

(7) After a snowfall, detection of the demolitions by the enemy is extremely difficult.

d. *Disadvantages.*

(1) Emplacing the explosives requires considerable time even when ice cutting equipment is available.

(2) The charges can be set off when hit by artillery fire.

(3) The gaps blown in the ice tend to freeze over rapidly in low temperatures.

e. *Tactical Use.* Ice demolitions are used for protection from frontal or flanking attacks. Normally, one or more sets of charges are laid close to the friendly shore and others farther out in the direction of the enemy (fig. 142). If desired, the enemy may be allowed to advance past the first set of charges and then both detonated at the same time. The enemy thus will be marooned on an ice floe, unable to continue to advance or retreat, and can be destroyed. The same trapping method may be used against enemy armor, or the charges may be detonated directly under the advancing tanks. Ice demolitions must be kept under observation and secured by friendly fire.

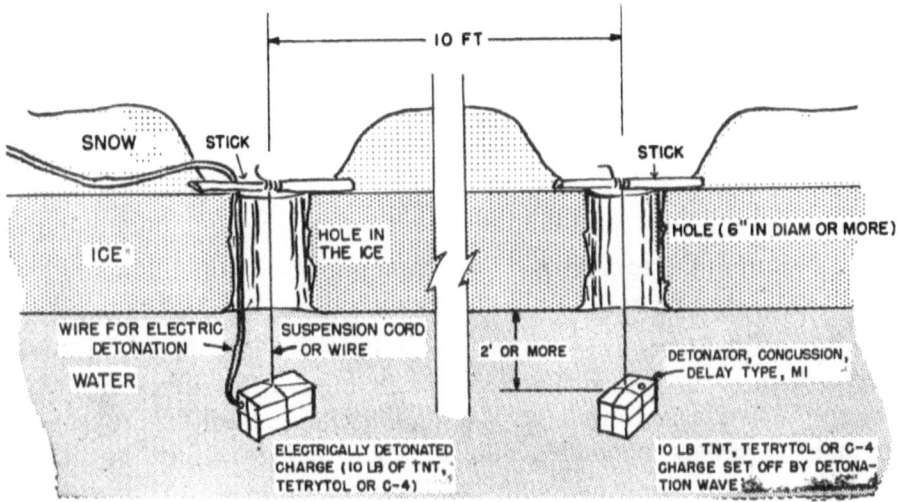

Figure 140. Method of placing charges in ice.

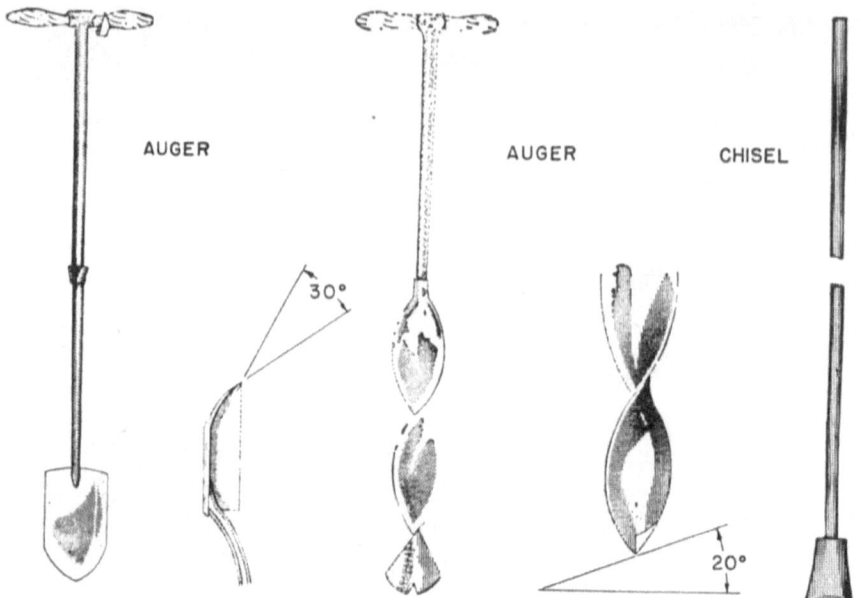

Figure 141. Types of ice augers and ice chisel.

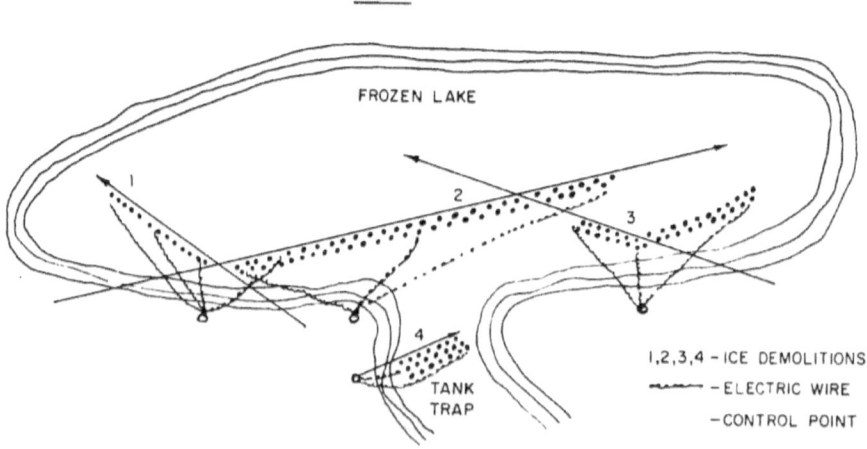

Figure 142. Ice demolitions.

179. Natural Obstacles

a. Snow-Covered and Icy Slopes. A steep slope is an obstacle to troops and vehicles even under normal conditions. When covered by deep snow or ice, it becomes much harder to surmount. The bogging-down action and the loss of traction caused by deep snow frequently create obstacles out of slopes which might be easily overcome otherwise. Pads of track-laying vehicles should be removed when encountering this type of terrain.

b. Windfalls. Occasionally, strong winds knock down many trees in a wooded area. These fallen trees are known as windfalls. They are very effective obstacles when covered with snow, especially to personnel wearing skis or snowshoes.

c. Lakes and Streams. Not all natural obstacles are equally effective in the winter as in the summer. Normally, bodies of water are considered natural obstacles, but under winter conditions the ice which forms may turn these former obstacles into excellent avenues of approach. This illustrates an important reason for reevaluating defensive positions before cold weather arrives.

d. Avalanches. An avalanche makes an excellent obstacle for blocking passes and roads. Since it occurs in mountainous country where there are few natural avenues of approach, an avalanche can have a far-reaching influence over combat operations. The problem with those avalanches which occur naturally is that, unless their timing and location are just right, they may be of help to the enemy. It is possible to predict in advance where an avalanche can and probably will occur. Then by the use of recoilless rifle or artillery fire, bombs, or explosives it is possible to induce the avalanche to slide at the desired time. This type of avalanche is an artificial obstacle in the technical sense. Generally it will be of more value than the natural type. Precautions against avalanche hazard are covered in FM 31-72.

180. Artificial Obstacles

a. Barbed Wire. There are many types of artificial obstacles used under summer conditions which are appropriate for winter use. Barbed wire normally employed makes an effective obstacle in soft, shallow snow. Triple concertina is especially effective since it is easy to install in addition to being difficult to cross. As the snow becomes deeper and more compacted, a point is reached where it is possible to cross the barbed wire on top of the snow. One type of barbed wire obstacle built to overcome this problem is known as the Lapland fence (fig. 143). Types of wire entanglements and winter obstacles are covered in FM 5-15.

b. Lapland Fence. The Lapland fence uses a floating type of anchor point or one which is not sunk into the ground. Poles are used to form a tripod. The tripod is mounted on a triangular base of wood. Six strands of wire are strung along the enemy side of the fence, four strands along the friendly side, and four strands along the base. As the snow becomes deeper, the tripods are raised out of the snow with poles or by other means to rest the obstacle on top of newly fallen snow. The base of the tripod and the base wires give enough bearing surface to prevent the fence from sinking into the snow.

c. Abatis. An abatis is similar to a windfall. Trees are felled at an angle of about 45 degrees to the enemy's direction of approach. The trees should be left attached to the stump to retard removal. Along trails, roads, and slopes, abatis can cause much trouble for skiers and vehicles.

d. Iced Road Grades. A useful obstacle can be made by pouring water on road grades. The ice that forms will seriously hamper vehicular traffic.

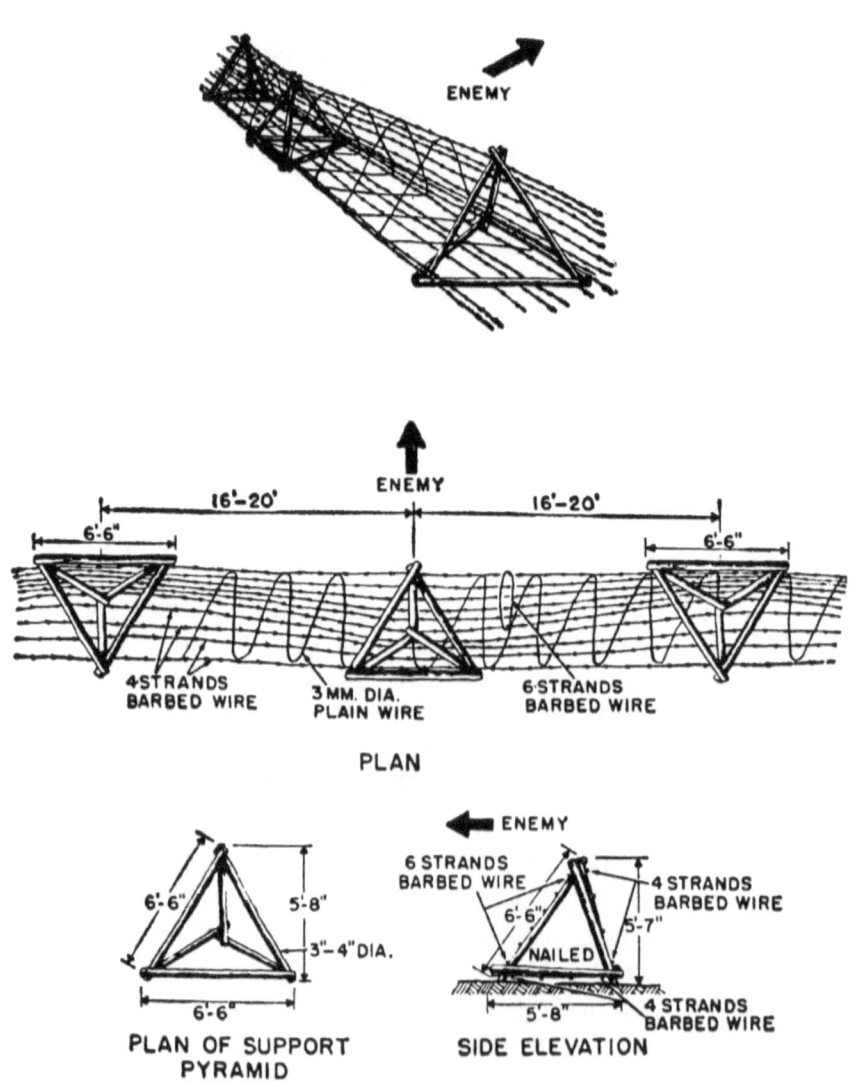

Figure 143. Lapland barbed wire.

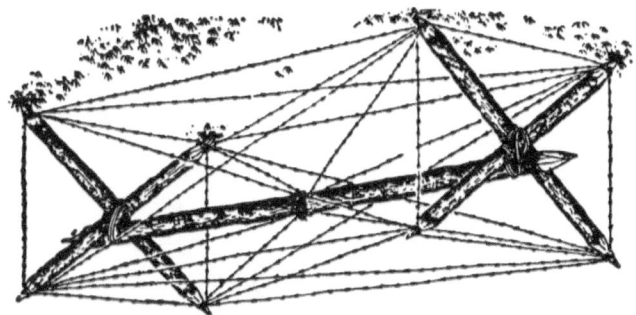

Figure 144. Knife rest.

181. Means of Improving Obstacles After Heavy Snowfalls

a. Knife Rests. Knife rests are portable barbed wire fences, usually constructed prior to the snowfall. The fences are constructed by tying two wood poles at their center, forming an "X". A similar "X" is made out of two other poles and then the two "X's" are lashed at either end of a 10 to 12 foot pole. This forms a framework to which barbed wire is fastened on all four sides. The obstacle can be stored until needed and then easily transported to the desired location.

b. Concertina Wire. Concertina wire is another quick way to improve on snow-covered obstacles. The concertina comes in 50 foot sections which can be quickly anchored to the top of existing obstacles.

c. Additional Barbed Wire. The possibility of using additional barbed wire strands should not be overlooked. Frequently, obstacles will have protruding poles to which extra barbed wire strands can be tied. Also, additional strands placed underneath such floating obstacles as Lapland fences and knife rests will help prevent the enemy from tunneling under these obstacles.

CHAPTER 7
SMALL UNIT LEADERS

Section I. GENERAL

182. Leadership Traits

a. The traits, qualities, and abilities requisite to good leadership in any theater of operations assume their greatest importance during operations in cold weather areas. Leaders must be impressed with and made clearly aware of this fact. During training, those men found incapable of meeting the necessarily rigid standards should be immediately removed and their places filled by better qualified individuals.

b. Military leadership is the art of influencing and directing men to an assigned goal in such a way as to obtain their obedience, confidence, respect, and loyal cooperation. The individual who demonstrates the traits of a leader and applies the fundamental principles of leadership will be a successful leader of men in cold weather areas.

c. All leadership traits as outlined in FM 22–100 are of importance to the leader assigned to units operating in cold weather areas. Peculiar conditions of cold increase the necessity for certain traits to a marked degree. Traits of utmost importance to the leader are:

    (1) *Initiative.* The energy or aptitude displayed in the initiation of action, self reliance, enterprise and self-initiated activity must be an outstanding characteristic of leaders who are involved in such operations, especially when units may become isolated. This requirement is more pronounced in the North than in other theaters of operation. In all training of leaders, initiative and improvisation must be carefully encouraged.

    (2) *Endurance and mental and physical stamina.* Extremes of climate and the isolation of units increase the necessity for strong mental and physical endurance. These conditions may cause early physical and mental fatigue, but can be overcome by determination, forcefulness, and aggressiveness.

    (3) *Unselfishness.* This is exemplified by the leader who does not take advantage of a situation for personal gain or safety at the expense of the unit. The physically competent,

vigorous leader who can resist the natural desire of first providing for his own comfort will be a successful and respected leader of his unit.

183. Leadership Principles

As in leadership traits, all leadership principles as outlined in FM 22-100 apply to leaders directing operations in cold weather latitudes, with particular emphasis placed on the following:

a. Know the Job. Every leader must know thoroughly the job at hand. The leader's actions must demonstrate to his subordinates his capabilities as a leader and his genuine desire to accomplish the mission with a minimum of effort expended by the men. The leader should frequently visit isolated units in adverse weather and show the men that he is a member of the team. He must earn the respect of the men and the right to command by a thorough understanding of the technical and tactical aspect of the task.

b. Know the Men and Look Out for Their Welfare.

   (1) The small unit leader must know the mental and physical capabilities of each of his men. Knowing this, he will be able to utilize them effectively. As an example, a strong stable soldier should be matched in the "buddy system" to guide and assist a weaker soldier.

   (2) In isolated areas recreation facilities normally are not available. It will be the leader's responsibility to insure that, during periods of rest or off-duty hours, men are not allowed to become psychological casualties. A good leader will gainfully employ his men, but not run the risk of "hounding" them. The good leader will, with ingenuity, devise projects which will occupy their minds and at the same time improve their professional qualifications as soldiers during periods of inactivity in isolated places.

   (3) In cold weather areas the problem of obtaining supplies assumes major proportions. Supply economy must be enforced at all times. Clothing and equipment must be checked frequently and maintained in first class condition. Continuous individual supervision on the part of the leader is mandatory.

   (4) Under adverse conditions the standards of personal hygiene and group sanitation will gradually become lower if not carefully supervised. These lowered sanitation standards are a sure indication that supervision is lacking and that morale is slipping. Men must not be allowed to become lazy about their personal habits. Rules of per-

sonal hygiene and sanitation must be enforced by the leader at all times.

c. Insure That the Task is Understood, Supervised, and Accomplished. Orders issued must be well thought out. When required, the leader must be prepared to take the leading part in carrying them out. Issuing an order is only the first and relatively small part of the leader's responsibility. The principal responsibility lies in supervision to insure that the order is properly executed. Cold regions can be friendly, but at the same time do not allow for errors or carelessness. An effective commander leads, not drives, therefore he must be able to differentiate between the two.

Section II. PECULIAR PROBLEMS OF LEADERS

184. Mental Processes

a. Cocoon-Like Existence. Many men, when bundled up in successive layers of clothing and with the head covered by a parka hood, tend to withdraw within themselves and to assume what has been termed a "cocoon-like existence." When so clothed, an individual's hearing and field of vision are greatly restricted and he tends to become oblivious to his surroundings. His mental processes become sluggish and although he looks, he does not see. These symptoms must be recognized by leaders and overcome. The leader must realize that it can happen to him and must be alert to prevent the growth of lethargy within himself. He must always appear alert to his men and prevent them from sinking into a state of cocoon existence. The remedy is simple and basic: Activity. Throw the parka hood back and engage in physical activity. Although the remedy is simple, the recognition of the condition requires leadership.

b. Individual and Group Hibernation. This process is again a manifestation of withdrawal from the surrounding environment. It is generally recognized by a tendency of individuals to seek the comfort of sleeping bags, and by the group remaining in tents or other shelter at the neglect of their duties. In extreme cases, guard and security measures may be abandoned and the safety of the unit jeopardized. The remedy is simple: Activity. The leader must insure that all personnel remain alert and active. Rigid insistence upon proper execution of all military duties and the prompt and proper performance of the many group "chores" is essential.

c. Personal Contact and Communication. It is essential that each individual and group be kept informed of what is happening. Due to the normal deadening of the senses a man left alone may quickly become oblivious to his surroundings, lose his sense of direction and his concern for his unit, and in extreme cases, for himself. He may become like a sheep and merely follow along, not knowing nor caring whether his unit is advancing or withdrawing. Each commander must take strong measures to insure that each small unit leader keeps his subordinates informed. This is particularly true of the company commanders keeping their platoon leaders informed, of platoon leaders informing their squad leaders, and the squad leaders informing their men. General information is of value but greatest importance must be placed on matters of immediate concern and interest to the individual. The chain of command must be rigidly followed and leaders must see that no man is left uninformed as to his immediate surroundings and situation.

d. Decentralization. Due to wide dispersion of units, isolated small groups of platoon and squad size, tent group or squad living, and difficulty of movement, the need for decentralization of command is emphasized. Orders must be concise, complete, and clear. The overall plan must be understood and each leader permitted to use his initiative and ingenuity in accomplishing his mission. Each commander must appreciate the additional time required to accomplish tasks under conditions of extreme cold and be patient in awaiting results from his subordinate units. Time and patience are perhaps the keys to decentralization.

e. Conservation of Energy. Two environments must be overcome in cold regions; one created by the enemy, and the second created by the climate and terrain. The climatic environment must not be permitted to sap the energy of the unit to a point where it can no longer cope with the enemy. The leader must be in superior physical condition or he cannot expend the additional energy required by his concern for his unit and still have the necessary energy to lead and direct his unit in combat. He must remember that there are seldom any tired units, just tired commanders.

185. Summary

a. The leader who is selected to lead troops in areas of the world where the extreme cold and rugged, trackless terrain make living and fighting more difficult, will face one of the greatest challenges of his lifetime.

b. He must possess the highest qualities of leadership and have the initiative, the confidence, and the endurance to utilize these qualities to the utmost. He must have the woodsman's knowledge of bushcraft and be able to navigate over rugged, trackless terrain. He must be physically strong, mentally alert, and able to stand on his own two feet and make decisions when on independent missions.

c. He must be more proficient than others, not only in command but in actual doing. He must be able to improvise and to teach his men to do likewise. He must be able to endure greater hardships than his men and be quick to recognize indications of mental lethargy. He must know the weaknesses and strengths in his men so that he may pair them more effectively in the buddy system. He must be firm when issuing orders but must also realize that as the men become colder and more miserable the time required to accomplish a task will be greatly increased. He must have patience and understanding and be able to lead without driving. In short, he must be the prototype of all leaders.

d. Military operations can be carried out successfully under the extreme conditions and over the difficult terrain conditions peculiar to the cold areas of the world. The task of the troop leader under conditions such as these becomes more difficult, but not impossible.

e. The leader must face up to his responsibilities and expend unselfishly and tirelessly of his time and his talents toward the betterment of the safety, the welfare, and the morale of his men.

f. The troop leader who knows his job and who makes proper application of the principles of leadership will earn the confidence and respect of his men, will be successful in the accomplishment of his mission, and will discover that his tour of duty in these cold areas will be both interesting and rewarding.

APPENDIX I
REFERENCES

AR 220-50	Field Organizations: Regiments—General Provisions.
AR 220-59	Ordnance Service Within Major Commands.
AR 220-60	Field Organizations: Battalions—Battle Groups—Squadrons—General Provisions.
AR 220-70	Field Organizations: Companies—General Provisions.
AR 320-50	Authorized Abbreviations.
AR 385-55	Prevention of Motor Vehicle Accidents.
AR 700-8400-1	Issue and Sale of Personal Clothing.
SR 320-5-1	Dictionary of United States Army Terms.
FM 5-10	Routes of Communication.
FM 5-15	Field Fortifications.
FM 5-20	Camouflage, Basic Principles.
FM 5-25	Explosives and Demolitions.
FM 5-31	Use and Installation of Booby Traps.
FM 6-40	Field Artillery Gunnery.
FM 7-10	Rifle Company, Infantry and Airborne Division Battle Groups.
FM 9-5	Ordnance Service in the Field.
FM 20-15	Tents and Tent Pitching.
FM 20-32	Employment of Land Mines.
FM 21-5	Military Training.
FM 21-6	Techniques of Military Instruction.
FM 21-10	Military Sanitation.
FM 21-11	First Aid For Soldiers.
FM 21-15	Individual Clothing and Equipment.
FM 21-20	Physical Training.
FM 21-26	Map Reading.
FM 21-30	Military Symbols.
FM 21-76	Survival.
FM 22-5	Drills and Ceremonies.
FM 22-100	Command and Leadership for the Small Unit Leader.
FM 23-55	Browning Machine Guns, Cal. .30.
FM 31-71	Northern Operations.
FM 31-72	Mountain Operations.

FM 100-5	Field Service Regulations; Operations.
FM 100-10	Field Service Regulations; Administration.
TM 3-522-15	Mask, Protective, Field M9 and Mask, Protective, Field, M9A1.
TM 5-251	Army Airfields and Heliports.
TM 5-295	Military Water Supply.
TM 5-560	Arctic Construction.
TM 9-772	Cargo Carriers M29, M29c and T107.
TM 9-1940	Land Mines.
TM 9-1946	Demolition Materials.
TM 9-1950	Rockets.
TM 9-2002	3.5 inch Rocket Launchers M20 and M20B1.
TM 9-2810	Tactical Motor Vehicle Inspection and Preventive Maintenance Services.
TM 9-2855	Instruction Guide; Operation and Maintenance of Ordnance Materiel in Extreme Cold (0° to —65° F.).
TM 9-3058	Cal..50 Spotting Rifle M8; 106-mm Rifle M40. and 106-mm Rifle Mount M79.
TM 9-2300-203-12	Full Tracked Armored Personnel Carrier M59 (T59) and 4.2 inch Full Tracked-Self-Propelled Motor M84.
TM 9-8662	Fuel-Burning Heaters for Winterization Equipment.
TM 10-228	Fitting Foot Wear.
TM 10-275	Principles and Utilization of Cold Weather Clothing and Sleeping Equipment.
TM 10-530	Principles of Packing and Rigging Aerial Delivery Containers.
TM 10-703	Small Detachment Cooking.
TM 10-708	Outfit, Cooking, Small Detachment.
TM 10-725	Stove, Tent, M1941, Complete, and Burner, Oil Stove, Tent, M1941.
TM 10-730	Heaters, Tent, Gasoline, 250,000 BTU, Herman Nelson and Silent Glow.
TM 10-735	Stove, Yukon, M1950.
TM 21-305	Manual for Wheeled Vehicle Driver.
TM 21-306	Manual for the Full Track Vehicle Driver.
DA Pam 108-1	Index of Army Motion Pictures, Film Strips, Slides, and Phono-Recordings.
DA Pam 310-1	Index of Administrative Publications.
DA Pam 310-3	Index of Training Publications.
DA Pam 310-4	Index of Technical Manuals, Technical Regulations, Technical Bulletins, Supply Bulletins, Lubrication Orders and Modification Work Orders.

DA Pam 310-7	Index of Tables of Organization and Equipment, Tables of Organization, Type Tables of Distribution, and Tables of Allowances.
DA Pam 320-1	Dictionary of United States Army Terms for Joint Usage.
TB 3-205-2	Winterizing Kit, Protective Mask, M1.
TB SIG 189	Cold Weather Photography.
TB SIG 219	Operation of Signal Equipment at Low Temperatures.
TB MED 81	Cold Injury.
TC 55-5	Motorized Sled Trains.
DA ACP 135 (A)	Communications Instructions, Distress and Rescue Procedure.

APPENDIX II
MAINTENANCE AND OPERATIONAL PROCEDURES FOR VEHICLES IN COLD WEATHER

Section I. GENERAL

1. Scope

This appendix contains the maintenance and operational procedures commanders, mechanics, and drivers should understand to facilitate keeping vehicles operating during cold weather. It also contains hints and suggestions for safe driving and convoy operations that are necessary to accomplish missions during cold weather.

2. Purpose

The purpose of the following paragraphs is to explain exactly what must be done to reduce the adverse effects of cold weather on vehicles and the extra precautions that must be taken during winter driving. In order to more fully appreciate the importance of these procedures, a brief review of the effects cold weather has on vehicular parts and systems is needed. Instructions for operation and maintenance of ordnance materiel is covered in TM 9-2855.

Section II. EFFECTS OF COLD WEATHER ON VEHICLES

3. General

A great amount of effort and research has gone into giving the individual soldier the best clothes to keep him warm in cold weather. A vehicle is affected by cold in much the same manner as a man. Consider what would happen to a platoon of men if the platoon leader didn't take the necessary steps to compensate for the cold to which his men are exposed. The driver of a vehicle must realize the same effects of cold are suffered by motor vehicles and certain precautions are necessary.

4. Examples of Cold Weather Effects

a. Metals become brittle when cold. A chain which during warm weather would be capable of supporting the weight for towing a vehicle may break under the same strain during cold weather.

A slight blow from a hammer may cause a pin to shear or a hook to break.

b. Rubber, in warm weather, is flexible; during extreme cold it becomes stiff, and bending will cause it to break, e.g., when a vehicle is parked for several hours during sub-zero weather, flattened-out areas develop in tires; these flattened-out areas have little resiliency until after the tires have warmed up, incident to operation.

c. Water freezes and expands; while it is expanding in a restricted space (as in an engine) it has tremendous power, enough to crack the toughest of iron.

d. Canvas becomes stiff much the same as does rubber and it becomes difficult to fold or unfold without damaging it.

e. Glass, being a poor conductor of heat, will crack if it is exposed to any sudden increase in temperature.

f. Gasoline will not freeze but becomes more difficult to vaporize. Since only vapor will burn, combustion of gasoline inside an engine is more difficult and unburned gasoline dilutes the oil in the crankcase.

g. Oils have a tendency to become thick, and consequently retard the flow through the oil pump to places where it is needed for lubrication. Thickened oils also increase the drag on the entire engine, thus making it more difficult to turn over.

h. Grease, which is a semi-solid to begin with, becomes hard and loses a great amount of its lubricating properties.

i. Leather cracks unless properly treated with neat's foot oil.

j. Paint tends to crack very easily when exposed any great length of time. Linseed oil should be applied to rifle stocks and hand guards to prevent moisture from entering the grain causing split stocks and hand guards due to freezing.

Section III. PRODUCTS AND EQUIPMENT DEVELOPED TO IMPROVE SUBZERO VEHICLE OPERATIONS

5. Products

In order to improve operation of ordnance vehicles in extreme cold, special fuels, lubricants, and related items, and winterization equipment have been provided. Listed below are the principal special products and the use for which intended.

a. Oil, Engine, Subzero, (Symbol OES) is an engine oil that will remain fluid at —50° F., and affords adequate lubrication at temperatures from 0° to —65° F., also during engine starting and warm-up periods.

b. Lubricant, Gear, Subzero (Symbol GOS), for use in gear box and power train assemblies, will afford adequate lubrication and prevent channeling and metal to metal contact.

c. Grease, Automotive and Artillery (Symbol GAA) for wheel bearings and general application, is now authorized for temperatures ranging from —65° to 125° F.

d. Other special products for cold weather use are hydraulic brake fluid (Symbol HBA), shock absorber fluid (Symbol MIL 5606), and arctic-type antifreeze.

e. Alcohol in small quantities must be used in gasoline tanks to prevent condensation from freezing in fuel tanks and lines. Alcohol is also used in the special "Evaporator" kits of airbrake systems.

f. Complete listings of cold weather products for use in specific vehicles are contained in pertinent lubrication orders for the vehicle.

6. Winterization Equipment

Winterization equipment provided for vehicles to insure satisfactory starting, operation, and overnight storage in extreme cold consists of the following:

a. Personnel Heater Kit. This kit consists of an electrically-driven gasoline-burning heater, engine primer, and winter fronts. It is used to heat the crew compartment or cab and furnish warm air for the windshield defroster.

b. Power Plant Heater Kit. This kit provides an electrically-driven gasoline-burning heater which heats and circulates the coolant through the engine cooling system and a battery-heating element. This heater is designed to function during overnight halts when required by extreme cold, or for a prestarting engine warm-up period. It is the most important and most useful element among the special winterization items. Further information on operation and maintenance of personnel and powerplant heaters is contained in TM 9-8662 and on the Operating Instruction plate mounted on the dash of the vehicle.

c. Cold Starting Aid Kit (Slave Kit) M40. This is a selfcontained unit providing an auxiliary source of electrical energy and heat to aid in starting the engine and to warm vital parts of a cold vehicle. It may also be used to charge batteries in an emergency. For further information on the operation and use of the "Slave Kit M40" in starting and warming vehicles see TB Ord 390. For the unit motor pool, this kit is an indispensable troubleshooting device.

d. Closure Kits. These are of two types: to inclose the cab of the vehicle for protection of the crew, and to inclose the cargo compartment for protection of the passengers. When installed on the cargo compartment, an additional personnel heater is required for passenger comfort.

Section IV. COLD WEATHER MAINTENANCE

7. Vehicles

All maintenance outlined in appropriate TM's for a particular vehicle must be accomplished and extreme care taken to insure all adjustments are as exact as possible. Proper lubricants must be used and these can be readily determined by consulting the appropriate lubrication order. One loose battery terminal, points slightly out of adjustment, a sparkplug wire loose, a ground cable loose, or a frozen gas line are some of the deficiencies that can make starting vehicles difficult, or could prevent starting altogether.

8. Personnel and Techniques

a. A large portion of a man's energy, in arctic-weather areas, is expended in self-preservation. This, naturally, reduces the efficiency of personnel in the operation and maintenance of equipment. The efficiency is further reduced by the bulk and clumsiness of the clothing that must be worn in extreme-cold areas. As it is impossible to handle extremely cold metal with bare hands, some form of mitten or glove must be worn at all times. The resulting loss of the sense of touch further reduces the efficiency of personnel. Even the most routine operations, such as handling latches or opening engine enclosures, become exasperating and time consuming when they must be performed with mittened hands. Experiments have proven, for example, that the time required by men to screw a nut on the largest bolt available was twice as long when mittens were employed over a similar operation conducted with bare hands. The space required to insure access to controls, adjustable devices, and to assemblies which are commonly replaced or which require periodic adjustment, inspection, and cleaning is also increased when the bulky arctic clothing is worn. The measurements of personnel with warm weather clothing and arctic-weather clothing are compared below.

Clothing measurements

	Warm Weather	Arctic Weather
Hand (width)	4"	6"
Wrist (circumference)	7½"	21"
Head (circumference)	23"	38"
Breadth across shoulders	18"	32"
Foot (width & length)	3½" x 11"	5" x 14"

9. Drivers

Particularly during cold weather operation, drivers must exercise extreme care in starting and operating their vehicles. Improper starting methods can cause battery and starter failure. Improper operations and driving can cause major assemblies to become inoperative. The driver must:

a. Use correct starting procedures.

b. Report all deficiencies noted on the vehicle.

c. Keep vehicle properly maintained and allow time for the additional maintenance needed in cold weather as outlined in Section V.

Section V. COLD WEATHER STARTING

10. General

Improper starting procedures can permanently damage the engine cylinder walls, dilute the crankcase oil with gasoline, run down batteries, and burn out starters. Much of the difficulty in cold weather starting will be eliminated, however, when the driver applies correct procedures and makes proper use of starting aids.

11. Starting Aids

The following are some of the starting aids a driver has at his disposal:

a. Engine stand-by heater.

b. Primer pump.

c. Slave kit.

12. Powerplant (Engine) Heater

Vehicles expected to operate in cold weather for prolonged periods are equipped with a winterization kit. This consists of an engine heater, battery pads, battery box and insulation, hood and grill covers, and primer pump. The standby engine heater heats the coolant in the engine to near operating temperature, thus reducing the drag in the engine. It is located under the vehicle so that the heat generated in the heater is also deflected to heat the oil pan and the transmission.

a. Operating Instructions:

(1) Open coolant valves from block of engine to battery pads.

(2) Check for proper coolant level in cooling system.

(3) Open fuel valves on line leading to engine heater.

(4) Check vehicle for gasoline leaks. (Do not start a vehicle with a gas leak. Report it immediately.)

(5) Make sure all battery cables and ground cables are tight.
FOR ELECTRICALLY CONTROLLED HEATERS PROCEED AS OUTLINED IN 6 THROUGH 12 BELOW.
(6) Press PRESS TO TEST button on control box. If circuit to heater is complete, the lamp will light on control box.
(7) Start heater by holding heater switch in START position until light flashes on (approximately 2 min.).
(8) Move control switch to RUN position.
(9) Sound of combustion may be heard before light flashes on. However, do not move switch to RUN position until light is on.
(10) Electrically controlled heaters will cut down to a LOW FIRE automatically after engine has reached operating temperature. However, at extremely low temperatures or on large engines the heater may run on HIGH FIRE indefinitely.
(11) To stop heater, move heater switch to OFF position. Wait until light goes out (about 2 minutes).
(12) Shut off fuel and coolant valves.
FOR MANUALLY OPERATED HEATERS PROCEED AS OUTLINED IN 13 THROUGH 22 BELOW.
(13) Move operating lever handle to HIGH position.
(14) Wait (1-4 minutes) for fuel to reach top of burner wick.
(15) Excess fuel may have accumulated in the overflow tube; it will drip out after the handle is first moved to HIGH position.

Caution: **If fuel is still dripping after heater is ready for ignition (1–4 minutes), do not start the heater. Report it to your immediate supervisor.**

(16) Open air intake door.
(17) Move control switch to START position. Listen for blower motor. Combustion should occur within 30 seconds.
(18) On manually operated heaters the heat output must be manually reduced by moving operating lever handle downward to LOW position.
(19) To stop heater move operating lever downward to full OFF position. Wait until flame goes out (about 30 seconds).
(20) Move control switch to OFF position.
(21) Close air intake door.
(22) Shut off fuel and coolant valves.

b. Care and Maintenance for Safe Operations:
(1) Check entire system daily for any sign of gas leaks. Remember, never start a heater on a vehicle with a gas leak.

(2) Report difficult starting, low heat output, improper heat control, unusual noises, gas leaks, and any other troubles encountered during heater operation.
(3) Clean sediment bowl as required.
(4) Clean accumulated dirt from around heater.
(5) Check coolant line connections from heater to engine block for leaks.
(6) Remove burner assembly, inspect and clean it, and replace wick if necessary.
(7) Check for spare igniter and make sure it is in good condition.

13. The Engine Heater

With reasonable care and maintenance the engine heater will assist greatly in eliminating the adverse effects cold weather has on starting a vehicle. Repeated cold starts without some means of preheating a vehicle engine will cut down on engine life, cause starter to burn out, and cause batteries to be discharged. Sometimes it will not be possible to heat the entire engine, oil pan, and transmission to operating temperature before a vehicle must be ready for operation. However, even raising the temperature of the system 40 or 50° will increase the chances for better, more efficient vehicular operation. A guide to go by is that the engine heater on HIGH FIRE will heat a vehicle approximately one degree per minute. This depends on wind, how long the vehicle has been idle, and the amount of insulation you have on your vehicle.

14. Primer Pump

The primer pump is part of the winterization kit. It is used in conjunction with the engine heater to help vaporize the gasoline before it goes into the cylinders for combustion. Excessive use of the primer pump results in overrich mixtures, possibility of hydrostatic lock, and washdown of cylinder wall lubrication. Refer to the applicable TM for further information on specific type vehicles. If a hydrostatic lock is suspected and the engine will not turn over when the starter is engaged, remove the spark plugs and drain the cylinders in accordance with instructions contained in the TM concerning that particular vehicle.

15. Slave Kit

The slave kit should be used by a trained man. Unless a person is completely familiar with its operation it should not be used. The slave kit is designed to preheat vehicles by means of a long, flexible metal tube which diverts heat from the kit to the desired place on a vehicle.

a. Never start a slave kit unless the flexible tube is hooked up to heat outlet.

b. Use only the insulated handles for holding the flexible tube.

c. Never direct the heat from the kit to open gasoline or on vehicles that have gasoline leaks.

d. Plug in slave receptacles on slave kit in accordance with voltage of the vehicle (24 volts on all M series vehicles).

e. Preheat vehicle with slave kit or engine heater prior to slaving a vehicle from the slave kit or another vehicle.

16. Towing

Towing a vehicle to start it should be done only when all other means have been exhausted. Prior to towing, however, if at all possible the vehicle should be preheated. Serious damage will occur during towing if extreme caution is not exercised in this procedure.

a. Preheat engine if possible.

b. Make a complete before-operation check.

c. Be sure that all electrical connections are tight.

d. Be certain the brakes and steering are in good condition.

e. Never try to tow a hydromatic for starting purposes unless the transmission lever is in F–I HIGH RANGE ONLY.

f. Make sure there are no binding parts on the vehicle being towed.

g. Be absolutely certain that the driver of the vehicle being towed has clear vision to all sides.

h. If vehicle will not start within a reasonably short distance, determine the cause and have it corrected.

Section VI. CARE OF THE FIVE MAJOR SYSTEMS OF A VEHICLE

17. Vehicle Condition

Unless vehicles are kept in the best possible mechanical condition during cold weather they will not operate properly. Successful cold weather operation depends on a high standard of maintenance and proper starting procedures. A large portion of our deadlines can be attributed to too many cold starts and improper driving habits. The starting procedures and care and maintenance of starting aids have already been covered. Following are some helpful hints to operators on care of the five major systems of a vehicle (cooling, electrical, lubricating, fuel, and power train systems) in cold weather.

a. Cooling System:
   (1) Keep an antifreeze solution in the cooling system consistent with the lowest expected temperatures in the area of operations.
   (2) Arctic antifreeze comes ready-mixed. Never mix water or other types of antifreeze with Arctic antifreeze.
   (3) Keep system full at all times.
   (4) Keep all coolant connections leak-free.
   (5) Keep temperature gauge operative.
   (6) Use hood and radiator covers as required.
   (7) Make sure fan belt is correctly adjusted and in good condition.
   (8) Preheat cooling system before starting.

b. Electrical System:
   (1) Keep batteries fully charged.
   (2) Never allow any cell to go below 1240 specific gravity.
   (3) Keep all electrical connections tight, especially on battery terminals, ground cables, starter cables, and generator and regulatory cables.
   (4) Keep batteries free from corrosion, moisture, ice, and dirt at all times.
   (5) Always use correct starting procedures.
   (6) Keep insulation in battery box in good condition.
   (7) Keep battery heating pads free from leaks, and paint battery with acid resistant paint.
   (8) Never try to charge a battery unless the electrolyte temperature is at least 35° F.
   (9) Never try to charge a battery by running the vehicle engine for prolonged periods. The batteries will probably not be able to accept the charge due to the fact that the electrolyte temperature is not high enough.
   (10) Never add water to a cold battery or the water will remain on top of the acid and freeze. If water is added, start the vehicle and move it to allow water to mix with the electrolyte.
   (11) Never add acid to a battery as this will cause the lead plates to break up.
   (12) Check for and report bare wires.
   (13) Preheat batteries before starting engine.

c. Lubricating System:
   (1) Use correct lubricants in crankcase and all gear boxes.

(2) Keep all lubricants free from dirt and water.

(3) Check and replenish crankcase oil each time it is needed.

(4) Idle engine a little above normal to keep oil pressure up.

(5) Check for leaks.

(6) After draining oil from crankcase, clean out shroud pan around crankcase. Oil in the shroud pan is a fire hazard when the engine heater is operated. This is one of the reasons fires occur and is not the fault of the heater, but rather of negligent driver maintenance.

(7) Never idle the engine for long periods. Condensation and crankcase dilution will cause sludge to form in the crankcase.

(8) Preheat engine before starting.

d. *Fuel System:*

(1) Add one quart of denatured alcohol to a 30-50 gallon fuel tank at the time of filling. Addition of alcohol should be more or less than one quart, depending on the size of the fuel tank.

(2) Drain fuel tank sumps regularly to remove the alcohol-water precipitate from bottom of tank, then replenish with the correct amount of alcohol.

(3) Check for, report, and repair all gasoline leaks.

(4) Make sure the heat control valve is adjusted for winter operations.

(5) Use primer pump correctly.

(6) Never idle engine for long periods. This will greatly increase gasoline consumption and is hard on engines.

(7) Refuel vehicle after each operation when possible.

(8) Keep snow and ice away from fillercap while removing it and refueling vehicle.

(9) Use engine heater to preheat engine. This will aid in vaporizing gasoline.

(10) Tank carburetor icing will cause emission of dark exhaust smoke and reduction of rpm. Proper engine warm-up will prevent this. Carburetor icing, once it has occurred, may be eliminated by halting the tank, and "running up" the engine for about five minutes (tactical situation permitting). The engine should then be stopped and allowed to stand for about ten minutes. The icing condition should be eliminated when the engine is re-started.

e. *Power Train, Steering, Brakes, and Suspension System:*

(1) Keep U-joints, bolts, and nuts tight.

(2) Keep all mounting bolts tight.
(3) Keep steering system tight and well lubricated.
(4) Make sure brakes are adjusted properly.
(5) Keep vehicle well lubricated.
(6) Practice proper driving habits and sane driving over rough roads, snow, and ice.

Section VII. DRIVING IN COLD WEATHER

18. General

The basic rules for driving during cold weather include all of the rules that apply under normal conditions. However, the necessity to adhere to these rules with the increased hazards of ice and snow is magnified.

19. Visibility

Good all-around visibility is the first requirement for safe driving.

a. Remove all ice from all windows to give all-around vision.

b. Use defrosters to keep windshield free from ice.

c. Clean and adjust rear view mirror.

d. Use lights during snowstorms and just prior to dusk and dawn.

e. Allow for additional distance between vehicles when exhaust is causing ice fog.

f. Use a guide when backing up or where a guide can assist in picking a trail in deep snow.

20. Traction for Driving and Stopping

a. Use chains in deep snow and on ice. They will increase traction for both movement and stops.

b. Place brush or burlap under wheels to aid in movement through deep snow and on ice.

c. The correct method for applying brakes is especially important. Never jam on brakes as this will lock the wheels and cause the vehicle to skid and have a greater stopping distance. The correct method for braking a vehicle on snow and ice is to release accelerator slowly and apply brakes with a feathering action.

d. Keep pioneer tools on all vehicles ready for use in removing excess snow and for cutting brush.

21. Additional Hints for Safe Cold Weather Driving

a. Never sleep in a cab of a vehicle with engine or heater running. Exhaust gases may cause death by asphyxiation.

b. Always adjust speed to road conditions.

c. Keep proper interval and compensate for road conditions (three to eleven times greater stopping distance may be needed on snow and ice).

d. Slow down before going around a curve.

e. Make slow, steady turns and stops.

f. Keep windows open slightly when heaters are being used.

g. Never stop in the center of a road.

h. Never pull off to the side of a road unless the shoulder has been checked. Large ditches covered with snow give the appearance of a firm shoulder.

i. When hauling troops in the rear of a truck, be certain to instruct them to wait for the driver to assist in their off loading.

j. Never overcrowd the cab of a vehicle with extra personnel or extra equipment. This cramps the driver, cuts down on his vision, and prevents him from maneuvering freely.

k. During halts, always check the vehicle for any troubles which may have occurred during operation.

l. Remove frost from headlights and stoplights.

m. Above all, use good judgment, be alert for other drivers' errors, and obey all traffic rules and regulations.

APPENDIX III
GROUND/AIR EMERGENCY CODE FOR USE IN AIR/LAND RESCUE SEARCH

1. General

Experience has shown the requirement for simple visual signals for use in an emergency by personnel who have become lost, crashed, or parachuted (or who are members of search parties), and who have need for medical assistance, food, information regarding the route to be followed, etc. Three types of such visual signals are contained in figures 145, 146, and 147.

2. Visual Signals

The use of one or more of these signals or types of signals will depend on individual circumstances and availability of signal material. However, as far as possible, the following instructions will be adhered to with respect to the signals contained in figures 145 and 146:

a. Form signals by any available means. (Some of the means usually available in an emergency situation are strips of fabric, parachute material, pieces of wood, stones, boughs, or by marking the surface by tramping snow or staining with oil, etc.).

b. Make signals not less than 8 feet (2.5 meters) in length.

c. Take care to lay out signals exactly as depicted to avoid confusion with other symbols.

d. Provide as much color contrast as possible between material used and the background.

e. Make every effort to attract attention by other means such as radio, flares, smoke, or reflected light. Smoke is one of the best attraction methods, because it can be seen for a great distance and will be investigated by all pilots, both military and civilian, as a routine matter. Be sure to give your signal while the aircraft is approaching you. Do not wait until the aircraft is straight above or has passed by.

f. The emergency signals included in this manual should be reproduced for use by individuals and/or small units which are required to accomplish independent or semi-independent missions.

NO.	MESSAGE	CODE SYMBOL	NO.	MESSAGE	CODE SYMBOL
1	REQUIRE DOCTOR - SERIOUS INJURIES	I	10	WILL ATTEMPT TAKE-OFF	△
2	REQUIRE MEDICAL SUPPLIES	II	11	AIRCRAFT SERIOUSLY DAMAGED	□
3	UNABLE TO PROCEED	X	12	PROBABLY SAFE TO LAND HERE	◁
4	REQUIRE FOOD AND WATER	F	13	REQUIRE FUEL AND OIL	L
5	REQUIRE FIREARMS AND AMMUNITION	≫	14	ALL WELL	LL
6	REQUIRE MAP AND COMPASS	□	15	NO	N
7	REQUIRE SIGNAL LAMP WITH BATTERY AND RADIO	--	16	YES	Y
8	INDICATE DIRECTION TO PROCEED	K	17	NOT UNDERSTOOD	⊥
9	AM PROCEEDING IN THIS DIRECTION	←	18	REQUIRE ENGINEER	W

Figure 145. Ground/air visual signals for use in emergency by survivors.

NO	MESSAGE	CODE SYMBOL	NO	MESSAGE	CODE SYMBOL
1	OPERATION COMPLETED	LLL	5	HAVE DIVIDED INTO TWO GROUPS. EACH PROCEEDING IN DIRECTION INDICATED	 ↗↙
2	WE HAVE FOUND ALL PERSONNEL	LL	6	INFORMATION RECEIVED THAT AIRCRAFT IS IN THIS DIRECTION	↑↑
3	WE HAVE FOUND ONLY SOME PERSONNEL	++	7	NOTHING FOUND. WILL CONTINUE TO SEARCH	NN
4	WE ARE NOT ABLE TO CONTINUE. RETURNING TO BASE	XX			

Figure 146. Ground/air visual signals for use in emergency by search parties.

Figure 147. Ground/air visual body signals for use in emergency by survivors.

3. Conveying and Acknowledging Information

a. When it is necessary for an aircraft to convey information to individuals who have become lost or isolated, or to search parties, and two-way radio is not available, the crew will, if practicable, convey the information by dropping a message or by dropping communication equipment that will enable direct contact to be established.

b. When a signal has been displayed and is understood, the pilot will acknowledge by dipping the aircraft's wings from side to side or by other prearranged signals.

c. When a signal has been displayed and is NOT understood, the pilot of the aircraft will so indicate by making a complete right turn or by other prearranged signals.

APPENDIX IV
SKI DRILL

Section I. INDIVIDUAL DRILL

1. General

Ski drill and ski training should be given concurrently. Ski drill is kept to the minimum necessary for assembly, organization, instruction, and speedy reaction to commands. Only those infantry drill movements in FM 22-5 which are easily performed on skis are used. If weapon is included, it is either carried across the back with the sling over the left shoulder, butt at the right side, or attached to the rucksack, if used. Before falling in for drill, skis are strapped with running surfaces together, tip to tip, using one strap to secure them tightly together between the toe and heel section of the bindings. Poles are interlaced by drawing the shaft of one through the basket of the other.

2. Fall In

The command is FALL IN. A normal interval (40 inches) is taken and skis are held in the position of Order Skis.

3. Order Skis

(fig. 148)

This is the position of attention with skis, except during Inspection of Skis. The skis are grasped with the right hand between the toe and heel section of the binding and held in a vertical position with the edges to the front. The tips of the skis rest on the ground, on line with and touching the toe of the right boot. The poles are held by placing the left hand through both wrist straps and grasping both handgrips. They are placed in a vertical position with the baskets on line with and touching the left boot. Both elbows are held close to the body.

4. At Ease and Rest

The same procedure is followed as in FM 22-5, except that the skis take the place of the rifle.

5. Facings

Facings are executed as prescribed in FM 22-5, except that

Figure 148. Position of Order Skis.

the skis take the place of the rifle. The ski poles are held in the left hand.

6. Hand Salute, Dismounted

At the position of Order or Right Shoulder Skis, the salute is rendered in the same manner as the rifle salute. To accomplish this, release the grip on the ski poles with the left hand, allowing the poles to hang from the wrist while the salute is executed. Regrasp the pole handles after execution of the salute.

7. Right Shoulder Skis
   (fig. 149)

This is a four count movement. Being at the position of Order

Skis, the command is RIGHT SHOULDER SKIS. At the command SKIS, the skis are lifted vertically until the upper right arm is horizontal. At the same time, the left hand grasps the skis over the front edges and approximately 12 inches below the front of the toe section of the binding. The ski poles remain on the left wrist as the movement is executed. (TWO) The right hand moves down and grasps the skis over the front edges, midway between the ski tips and the front of the toe section of the binding. (THREE) Skis are lowered so that the balance point rests on the shoulder and the skis are at an angle of approximately 45° to the horizontal, with the right elbow close to the side. (FOUR) The left arm is cut smartly back to the side and the grip on the ski poles resumed.

Figure 149. Position of right shoulder skis.

8. Order Skis from Right Shoulder Skis

This is a four count movement. The command is ORDER SKIS. At the command SKIS, the left hand grasps the skis midway between the toe section of the binding and the right hand. The poles hang from the wrist by the straps. (TWO) The skis are brought down until they are in a vertical position approximately 18 inches from the ground. (THREE) The right hand grasps the skis over the rear edges between the toe and heel section of the bindings. (FOUR) The skis are lowered gently to the ground. At the same time, the left hand grasps the handgrips of the poles and is brought back to the left side.

9. Open and Close Ranks

Same as FM 22–5 except that each rank takes double the distance. For example:

a. Front rank takes 4 steps forward.

b. Second rank takes 2 steps forward.

c. Third rank stands fast.

d. Fourth rank takes 4 steps backwards.

10. Inspection Skis

(fig. 150)

Being at Order Skis, the command is INSPECTION SKIS. At the command SKIS, the skis are unstrapped with the left hand and the loose strap placed in the pocket. The skis are separated and the position of attention assumed, holding one ski in each hand between the toe and heel section of the bindings, running surfaces to the front and the tip of each ski in line with and approximately 3 inches outside the toe of the corresponding boot. The ski poles hang from the left wrist. After the inspecting officer has examined the running surfaces, the skis are rotated 180° to display the top surface. When the inspecting officer has passed, the skis are refastened and the position of Order Skis resumed.

11. Ground Skis

This movement is done in 3 counts. Being at Order Skis, the command is GROUND SKIS. At the command SKIS, take two steps to the rear, leaving the ski tips in place. Lower the skis partially to the ground by sliding the right hand toward the heels of the skis. (TWO) The skis are then placed on the ground, on edge. (THREE) A position is taken directly to the left of the ski bindings, facing the tips of the skis. The poles are placed

Figure 150. Position of inspection skis.

on the left, parallel to the skis, baskets to the rear, mid point of the shafts even with and close to the left boot. The position of attention is then assumed.

12. Take Skis From Ground Skis

The command is TAKE SKIS. At the command SKIS secure ski poles, reverse the three movements of Ground Skis, and assume the position of Order Skis.

13. Stack Skis
(fig. 151)

This movement is done in four counts. The command is STACK SKIS. At the command SKIS, the first two movements of Ground

Skis are executed. (THREE) The ski poles are separated and the points placed in the ground on each side of the ski heels, approximately 3 feet apart. A V is made of the handgrips by interlacing each wrist strap over the opposite handgrip and crossing the right pole in front of the left. The poles are then grasped with the right hand at the point where they intersect. (FOUR) The heels of the skis are picked up with the left hand and placed, edges up, running surfaces together, into the V-formed by the handgrips. At the same time, the poles are tilted forward so that they are approximately 18 inches from the ski heels. This increases support. A position of attention is then assumed beside the ski bindings, with the stack to the right.

14. Take Skis from Stack Skis

The command is TAKE SKIS. At the command SKIS, the movements of Stack Skis are reversed and the position of Order Skis assumed.

15. Stack Equipment

With skis stacked, the command is STACK EQUIPMENT.

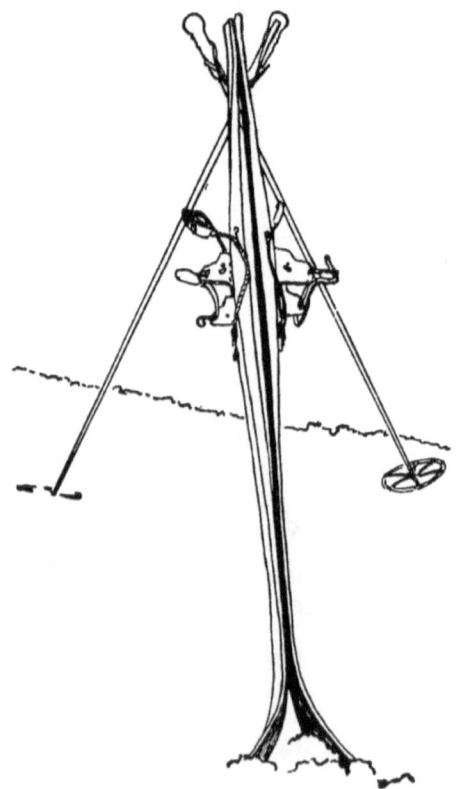

Figure 151. Position of skis in stack skis.

At the command EQUIPMENT, the pack is hung over the heels of the skis by both shoulder straps with the body of the pack to the right of the individual as he faces to the rear. The rifle remains attached to the pack when applicable or, when unattached, it is also hung to the right, vertically and with the receiver down. Any additional equipment is hung in a similar manner.

16. Take Equipment

The command is TAKE EQUIPMENT. At the command EQUIPMENT, the rifle, pack, and equipment are removed and a position of attention resumed beside the skis.

17. Mount Skis

This movement is done in 5 counts. Being in line at open ranks, the command is MOUNT SKIS. At the command SKIS, the first three movements of Ground Skis are executed. (FOUR) The skis are straddled. (FIVE) The skis are separated (from each other) placed on the ground and the boots are secured to the bindings. Poles are then separated and grasped with the left hand. The right hand is inserted up through the wrist strap from underneath so that the wrist strap is around the back of the wrist. Then the handgrip is grasped. This procedure is repeated with the left hand and the position of attention is assumed. On skis, this is as follows (fig. 152):

a. Skis are parallel and 3 to 4 inches apart, with the weight of the body evenly on both skis.

b. Poles are placed vertically with each basket in line with, and touching the toe of, the corresponding boot.

c. Elbows are close to the body, with the position of the hands dependent on the length of the pole.

18. Dismount Skis

The command is DISMOUNT SKIS. At the command SKIS, the movements of Mount Skis are reversed and the position of Order Skis assumed.

19. At Ease and Rest, Skis Mounted

When mounted on skis, the right ski must be left in place when At Ease is given. At Rest, both skis may be moved.

20. Hand Salute, Skis Mounted
(fig. 153)

When mounted on skis, the hand salute is rendered the same as prescribed in FM 22-5. The right hand is removed from the

Figure 152. Position of attention, skis mounted.

wrist strap if time permits. If time does not permit, the pole hangs from the wrist by the strap until after the salute is executed.

21. Ski Interval
(fig. 154)

Maneuvers on skis are done at ski interval. If skis are already mounted, ski interval will be taken by each individual when falling in, unless otherwise specified. The ski interval is approximately nine feet and is measured by extending both the right arm and right ski pole and the left arm with the left pole hanging from the wrist. When skis are mounted in ranks while at normal interval, the command is TAKE SKI INTERVAL TO THE RIGHT (LEFT). On this command, interval is taken as described in FM 22–5, except that the step turn to the left is executed rather than a face to the left in marching. The interval is measured as described above. If it is desired to straighten the ranks after ski interval has been taken, the command AT SKI INTERVAL, DRESS RIGHT DRESS is given.

Figure 153. The hand salute, skis mounted.

Figure 154. Taking ski interval.

22. Right or Left Face

When mounted on skis, this movement is executed in four counts. The command is RIGHT FACE. At the command FACE, the right ski is raised slightly and rotated 45° to the right, using its heel as a pivot. (TWO) The left ski is moved alongside the right ski. (THREE) The first movement is repeated. (FOUR) The second movement is repeated. Each ski pole is raised, moved, and placed with the corresponding ski. Left Face is executed in the same manner except the 45° movement is made to the left with the left ski.

23. About Face

This movement is executed in four counts. The command is ABOUT FACE. At the command FACE, the left pole is placed alongside the left ski approximately 18 to 24 inches in front of the toe. At the same time, the right pole is placed alongside the right ski approximately 18 to 24 inches in rear of the toe. (TWO) The right ski is raised until it is perpendicular, with its heel alongside the tip of the left ski. (THREE) Using the heel as a pivot, the right ski is rotated and placed alongside the right pole and pointing in the opposite direction. (FOUR) The left ski and ski pole are then brought around simultaneously and the left ski placed in the new direction alongside the right ski, with the left ski pole placed by the toe of the left foot.

Section II. UNIT DRILL

24. Moving at Right Shoulder Skis

a. Drill. To move men out at right shoulder skis, the preparatory command FORWARD is given with sufficient pause before the command of execution MARCH to allow the men to bring their poles up onto the left shoulder and placed with baskets to the rear under the skis (1, fig. 155). To move from Right Shoulder Skis to Order Skis after halting, the preparatory command ORDER is given with sufficient pause before the command of execution SKIS to allow the men to bring their poles down to the left side.

b. Marches. There are three methods of carrying skis which may be used in marching, their use depending on the length of march and the type of terrain. If the march is relatively short, at the command ROUTE STEP the poles may be removed from under the skis and brought down to the side at the discretion of the individual. This enables him to rest or warm this arm and hand or to use the poles for support when climbing a slope. Skis may also be alternately shifted from shoulder to shoulder to reduce fatigue. At the command SQUAD, PLATOON, or COMPANY ATTENTION, the position of Right Shoulder Skis is resumed with poles under the skis. Allow sufficient time between the preparatory command and the command of execution for individuals to place the skis and poles in proper position. For longer marches where the terrain is flat or rolling, the poles may be strapped to the skis with the baskets over the tips and the skis alternated between the right and left shoulder to avoid fatigue (2, fig. 155). This method is valuable in cold weather, as it enables the individual to alternate warming of each hand by swinging it or placing it under his outer clothing. For longer marches, especially over steep terrain, the skis may be tied to the rucksack. One is tied on each side with the tips up and strapped together at the top to form an A shape (3, fig. 155). This method allows the individual to use the poles for additional support either together, in one hand, or separately, one in each hand.

1   Using the ski poles for support.

Skis and poles strapped together to leave hands free.

3   A-frame method.

Figure 155. Various methods of carrying skis.

25. Flanking Movement From Normal Interval

This movement is used when it is desired to move men to the flanks when mounted on skis at normal interval. The movement is done in 4 counts. The commands are to the RIGHT (LEFT) FLANK AT INTERVAL, MARCH. At the command MARCH, the right flank man pivots on the heel of his right ski 45° to the right and slides slightly forward on it. (TWO) The left ski is brought up parallel to the right ski, allowing this ski also to slide slightly forward. (THREE) The first movement is repeated. (FOUR) The second movement is repeated and a normal pace taken in the new direction. As the right flank man takes his third step, the next man starts his first step. This procedure is followed by each man in line. When the last man has finished this movement, the unit will be marching in the new direction at ski intervals. Flanking to the left is executed in the same manner, except that the left flank man starts the movement with his left foot.

26. Flanking Movement From Ski Interval

This movement is made by first commanding RIGHT or LEFT FACE. When this facing has been completed the command FORWARD MARCH is given.

27. Column Movement

When mounted on skis, the commands are COLUMN RIGHT (LEFT), MARCH. At the command MARCH, the leading man takes a full step forward, then turns as in facing on skis, except that at each step a short slide forward is made. The fourth step is of full length in the new direction. Succeeding men follow in his trace. For COLUMN HALF RIGHT (HALF LEFT) the same procedure is followed except that the second step is of full length in the new direction.

28. To March to the Rear

For this movement, when mounted on skis, three separate commands are given, allowing each movement to be completed before the next command is given. These commands in order are HALT, ABOUT FACE, FORWARD MARCH. The about face is executed as described in paragraph 23.

APPENDIX V
EFFECTS OF COLD WEATHER ON INFANTRY WEAPONS

1. General

In cold areas many climatic conditions will greatly affect the operation and employment of infantry weapons. All individuals must be well aware of these conditions in order that they may properly handle and care for their weapons under these adverse circumstances.

2. Factors Affecting Weapons

a. Sluggishness. The most important and common problem is the sluggishness of the operation of the weapons in extreme cold. Normal lubricants thicken in low temperature and stoppage or sluggish action of firearms results. During the winter, weapons must be stripped completely and cleaned with a dry-cleaning solvent to remove all lubricants and rust prevention compounds. The prescribed application of Lubrication Oil, Weapons (LAW) should then be made. These lubricants will provide proper lubrication during the winter and help minimize snow and ice from freezing on the weapons.

b. Breakages and Malfunctions. Another problem that faces the soldier in the areas of severe cold is a higher rate of breakage and malfunctions. These can also be attributed primarily to the cold, although snow in a weapon may cause stoppage and malfunctions. The tempered metal of automatic weapons, for example, will cool to a point where it cannot be touched by human flesh. This extreme cold makes the metal brittle. When the weapon is fired at subzero temperatures, the temperature of the barrel and gun will rapidly rise to between 200 and 700°, depending upon the number of rounds fired. This again reduces the temper and, because the parts are working, breakages will occur early in the firing while the weapon is warming up. Many malfunctions also occur during this period due to the presence of ice or snow in the weapon or freezing of working parts. The weapons should first be fired at a slow rate of fire. Once the parts have warmed up, the rate of fire may be increased. One of the main problems is to insure that snow and ice do not get into the working parts, sights, or barrel. The weapon must be carefully handled during movement through the snow-covered woods, and especially under combat conditions in deep snow. In the bivouac area a rifle stand (fig. 39) should always be constructed to protect the weapons from the elements.

c. Condensation. Condensation forms on weapons when they are taken from the extreme cold into any type of heated shelter. This condensation is often referred to as "sweating." When the weapon is taken out into the cold air, the film of condensation freezes, especially in the internal parts, and stoppages and malfunctions result. For this reason the weapons must be left outdoors or stored in unheated shelters. When weapons are taken into heated shelter for cleaning purposes, "sweating" may continue for as long as 1 hour. Therefore, when time is available, wait 1 hour, remove all condensation, and then clean the weapon. In addition, this action will prevent corrosion.

d. Ice Fog. A problem of visibility close to the ground occurs when an automatic or heavier weapon is fired in temperatures below —20°. As the round leaves the weapon, the water vapor in the air is crystallized, creating minute ice particles which produce ice fog. This fog will hang over the weapon and follow the path of the projectile, obscuring the gunner's vision along his line of fire. If the air is still, the ice fog will remain for many minutes and hover in one place. Therefore, the weapon will have to be displaced to the right or left to again secure use of its sights if firing is to be continued.

e. Emplacement. Most crew-served infantry weapons need a natural "base" or gun platform to raise the barrel above the snow surface so it may be fired accurately. In summer the ground provides a solid base and yet has enough resilience to act as a shock absorber. In winter the soft snow gives under the fire of the gun. If the weapon is emplaced on the solid frozen ground, there is no "give" and all the shock of firing is absorbed by the weapon itself, resulting in breakage (par. 156). Also the slippery surface of the frozen ground may allow the weapon to slide. If the snow is not too deep, and if time is available, tripods and baseplates should be dug into the ground or solidly positioned by expedient means to keep them from moving.

3. Cold Effects on Various Types of Infantry Weapons

a. Pistol, Automatic, Cal .45. The Pistol, Automatic, Cal .45, cannot be fired while one is wearing arctic mittens. The firer must remove his mittens or use the lighter weight trigger finger mitten. The only other difficulty that may be encountered is the breakage of moving parts due to the cold. However, this is not as serious as with the larger automatic weapons.

b. Rifle, US, Cal .30, Carbine, Cal .30 and Rifle, Automatic, Cal .30. Firing of these weapons will also necessitate the use of trigger finger mittens. This means that the individual cannot operate the weapon over a sustained period of time in extreme cold temperature. All these weapons create ice fog. However, since the firer can readily move his position, this poses no problem. The main problem is that more malfunctions and breakages are caused in firing due to the cold or fouling of the weapons with snow or ice. The carbine is sensitive and also the most difficult to repair because the parts of the weapon are small and difficult to handle with gloves or mittens on. In addition, the M-1 Rifle has the problem of the gas port locking screw freezing. Parts most subject to breakage are sears, firing pins, and operating rods—parts that are moving and are affected by recoil. Malfunctions of the automatic rifle may be caused by defective and plugged magazines. Special care must be taken to keep magazines free from snow. Also, the wing nuts on the bipod, bipod parts, and bipod ring tend to freeze together or to the barrel of the weapon. To avoid this problem, apply Lubricating Oil, Weapons (LAW) on the part concerned.

c. Gun, Machine, Cal .30 and Cal .50. These weapons should be well lubricated because of their many moving parts. If Lubricating Oil, Weapons (LAW) is not available, these weapons, when fired cold and dry, will have fewer malfunctions if fired at a slow rate of fire. Once the parts have warmed up, temperate lubricants may be applied and the rate of fire may be gradually increased. MG's have a high rate of breakages and malfunctions due to the cold weather. Parts especially affected are sears and bolt parts. Extra parts of this type should be carried by the gunner and his crew. The most common malfunction, occurring early in the firing, is called short recoil (bolt does not recoil fully to the rear). Applying immediate action will, as described in FM 23-15, reduce this stoppage and, as the metal warms, the problem will diminish. A second malfunction is caused by freezing and hardening of the buffer. This, in turn, causes great shock and rapid recoil, thereby increasing the cyclic rate of the weapon. When this happens and the gun continues to fire, something has to give, and generally parts will break. Condensation will cause the freezing of parts, as on most other weapons. Ice fog greatly impairs the operation. Therefore, 2 to 3 alternate gun positions must be prepared.

d. Gun, Submachine, Cal .45. This weapon has problems similar to the other small arms, especially with condensation, because it is normally carried inside a warm vehicle. The magazine should be well protected from snow.

e. Grenade, Hand. No particular problems exist in the use of hand grenades in extremes of climate, with the exception that they lose much of their effectiveness when detonated under snow. The following are precautions necessary for throwing hand grenades by personnel wearing arctic handgear under extreme cold conditions.

    (1) Handgear must be completely dry. Handling of snow and ice with gloves or mittens may result in grenades freezing to the wet handgear.

    (2) Grenades must be held near the neck of the fuze to avoid slipping or turning of the grenades when the safety pins are removed.

    (3) Right hand throwers must place the grenade so that the safety lever rests on the first knuckle of the thumb to insure a sensitive feeling of the safety lever.

    (4) Left handed throwers must place the grenade so that the safety lever rests between the first and second knuckles of the fingers, to insure a sensitive feeling of the safety lever and good access to the safety pin ring.

f. Launcher, Grenade, Rifle. No special problems have been encountered with the launcher except that the range may be slightly reduced due to the slower burning of the crimped cartridge. As a result, the sight may not hold true and may have to be adjusted.

g. Launcher, Rocket, 3.5-Inch. The main problem centers around the ammunition used. The 3.5-inch round has a burning propellant which moves it toward the target. This propellant, due to the effect of the blast and its slow burning quality in cold weather, is highly dangerous at low temperatures. The gunner and loader can be burned and lacerated by particles of the burning propellant which are thrown back as the round leaves the muzzle of the launcher. The firing of this weapon (under a peacetime situation) is normally restricted to temperatures above —20° F., but extreme caution should be exercised when firing at any temperature below freezing. Both gunner and loader must be equipped with face masks and gloves. Also, due to the characteristics of cold, dry air the back blast danger area, which is designated in TM 9–2002 for temperate climates as a triangular area with a base and height of 25 yards, the apex of the triangle being at the breech of the launcher and its height (which bisects the base) an extension of the launcher axis, is greatly increased, and precaution should be exercised in operating behind the weapon. It has no emplacement problems, but will create ice fog and will have to be moved often to prevent detection when the fog persists. Also, the range is reduced due to the slow-burning propellant. The gunner will have to make his own firing tables and probably sight slightly high, especially at longer ranges.

h. Rifle, 106mm. Since there is no burning propellant in the projectile itself, the crew is quite safe from flying particles. Due to the slow burning qualities of the propellant, however, the firing data for temperate climates cannot be used and the weapon must be zeroed for the temperature in which it is being fired. Once zeroed, the weapon is highly accurate. The back blast danger area in winter is almost triple that of the back blast area of the weapon in temperate climates; for example, TM 9-3058 states in part: "The rear blast of this weapon is dangerous and must always be kept in mind. The danger zone for personnel (*when firing in temperate climates*) extends at least 100 feet to the rear of the weapon and spreads out to about 175 feet each side of the center line of the weapon. Windows and structures at least 300 feet to the rear of the weapon may be damaged due to the air pressure." These distances must be *tripled* to provide minimum safety when firing the weapon under conditions of extreme cold. In cold weather, when the propellant burns slowly, the rate of fire will be slower because, after the round leaves the muzzle, burning gases remain in the barrel and the weapon cannot be reloaded until they burn out. This phenomenon is known as "afterburn." Gunners should exercise extreme care to avoid premature explosion of the round in the weapon. A period of at least 60 seconds must elapse between firing and reloading.

> (1) One of the major problems in the firing of the recoilless rifle is the ice fog effect, requiring frequent displacement of weapons to avoid detection. The Spotter-Tracer Rifle, Cal .50 M-8c, atop the weapon creates problems due to the fact that its trajectory and that of the recoilless round do not coincide. The spotter-tracer rifle is also subject to breakage of metal parts, such as the firing pins, which become brittle due to the cold. The working parts should be lubricated with Lubricating Oil, Weapons (LAW). Otherwise the weapon is best fired dry.
>
> (2) Another phenomenon occurs in extreme cold affecting recoilless rifles. The problem is a deformity of the barrel due to solar radiation, or the heating of the atmosphere and ground. This will happen if the weapon is boresighted, for example, prior to sunrise. If the sight reticle and the bore have been placed on the same target in the early morning hours, after the sun rises the bore

may be pointing at one target and the sight reticle at the original one. The barrel has actually bent slightly due to the increase in temperature and thus the zero has been lost. After the weapon has been fired for several rounds, it is again boresighted and retains its high accuracy. Therefore, this phenomenon concerns itself primarily with gaining first round hits; the gun crew must be aware of this condition and how to correct it. The gunner of the 106-mm rifle should not trust the firing tables in low temperatures, but should make his own data for cold weather conditions.

i. Gun, 90mm, SP. This weapon has many of the problems of the Rifle, 106mm. Breakage and malfunctions are few. The two primary problems are the formation of ice fog when the weapon is fired and distortion of the tube caused by solar radiation. The gunner is required to recompute his firing data as stated above.

j. Tank, 76mm Gun. The problems of lubrication and breakage are greatly diminished due to the fact that most of the working parts of the weapon are inclosed in a warmed turret. The major problem is the effect of temperature changes on the ammunition. Ammunition stored inside the turret will be warm and have the same general ballistic characteristics of ammunition fired in temperate climates. The weapon is probably zeroed with this warm ammunition. Other ammunition is stored outside the tank where the temperature is extremely cold. When this ammunition is fired, the powder will burn slowly and it will have completely different ballistic characteristics, thus rendering the initial zero useless. If possible, the ammunition brought in from the outside should be heated in the turret before firing. In a combat situation this is not practical because the ammunition may have to be used immediately. The gunner must have his own data for cold ammunition or be ready to hastily rezero the weapon. In either case he will have to make sight adjustments. There is also a problem with solar radiation, ice fog, muzzle blast fog, and snow particles being blown up in front of sights and obscuring the visibility of the gunner.

k. Mortar, 81mm or 4.2-Inch. The matter of breakage in mortars is a minor one since there are few parts. However, firing pins often get brittle and break. The baseplate must be solidly positioned to prevent sliding. It may be necessary to dig into the ground to accomplish this. When the weapon is emplaced on frozen ground, the combination of the cold making the metal brittle and the tremendous shock that the baseplate receives when a round is fired, occasionally may cause the baseplate to crack. Frozen ground has no resiliency, and the baseplate and other bracing parts of the weapons absorb the entire shock of firing.

(1) One field expedient that will reduce the possibility of a cracked baseplate is to place a brush matting under the baseplate. The matting should be thick enough to act as a shock absorber, but not so thick as to cause the baseplate to bounce out of its dug in position. Another method of positioning the weapon is to place bags of dry sand or snow beneath the baseplate. The sandbags will provide the weapon with a solid, yet resilient, shock absorbing base. An additional problem with the mortars is that they cannot be handled without touching bare metal as can other infantry weapons with wooden or plastic handles and stock. The crew must keep their gloves or mittens on and avoid touching the metal surface with bare flesh. There are practically no lubrication or ice fog problems with the mortars. Malfunctions are also quite infrequent.

(2) The ammunition is affected by the cold in the same manner as the other types of ammunition. Firing tables may be utilized provided the proper range K's are established through experience. Consult FM 6-40 for charge restriction at low temperatures. The VT-fuze type ammo is considered the most effective mortar ammunition in the northern latitudes in the winter. Other contact-detonated ammunition will penetrate the snow before exploding and much of its effectiveness is lost and dissipated in the snow. A greater frequency of short rounds, as much as 1000 to 1500 yards short, may be experienced at low temperatures from the 4.2-inch mortar.

l. Rifle, M-14. (The light barreled version scheduled to replace the M-1 Rifle, Carbine, Submachine gun and Rifle, M-15.) (The heavy barreled version designed to replace the automatic rifle.) Tests under cold weather conditions have shown these weapons to be very satisfactory for use in extreme low temperatures. The weapons are found highly accurate and to have less breakages and malfunctions than any of the weapons they will replace.

m. Gun, Machine M-60. (Scheduled to replace the infantry's present family of machine guns.) Tests have proven it to be a highly effective weapon in cold weather areas. It was still subject to some breakage of recoiling parts, and occasionally the buffer froze and increased the cyclic rate of fire. The weapon is readily assembled or disassembled due to the larger parts which are much easier to handle with gloves on. The main problems with the gun occurred in visibility, due to ice fog, lubrication, which is corrected by the use of Lubricating Oil Weapons (LAW), and the malfunction of a short recoil with cold guns.

n. Portable Flame Thrower M2A1. From experience it has been found that the portable flame thrower would have limited use at extreme low temperatures. The rubber components, particularly the fuel hose, become rigid at temperatures below 0° F. Servicing of the flame thrower must be accomplished outside shelters to prevent frost from accumulating in the air pressure regulator. When firing the flame thrower at low temperatures, two or more ignition charges should be used to insure ignition of the fuel. The preparation of thickened flame thrower fuel at low temperatures according to prescribed mixtures is less dependable, and sample batches of prepared fuel should be test fired whenever the situation permits.

APPENDIX VI
WEIGHTS OF COLD WEATHER CLOTHING AND EQUIPMENT

1. General

The weights shown in this appendix are for standard issue items of individual clothing and equipment presently in the supply system. As new items of clothing and equipment are developed and standardized, their weights will be reflected by changes to the manual.

2. Purpose

This appendix gives a breakdown of weights by item of the individual loads illustrated in figures of this manual. Abbreviated nomenclature has been used in most instances.

3. Weights

Weights of various types of individual loads are shown in chapter 2. Weights of *items* that are carried under various weather conditions are shown below.

a. Combat load worn and carried in moderately cold weather (a, fig. 1).

    (1) Existence Load:

        (*a*) Clothing:

Quantity	Nomenclature	Weight
1 ea	Undershirt, man's (winter)	.94 lbs
1 ea	Drawers, men's (winter)	.94
1 ea	Trousers, men's, cotton, wind resistant, sateen, 9 oz wt, OG, QM shade 107, water repellent treated	2.25
1 ea	Liner, trousers, arctic, field	1.62
1 ea	Shirt, wool, OG 108	1.50
1 ea	Coat, single breasted (field jacket)	3.25
1 ea	Liner, mohair frieze (field jacket liner)	2.06
1 ea	Hood, winter, OG, w/fur ruff	.75
1 ea	Cap, pile	.44
1 ea	Suspenders, trouser	.25
1 pr	Mitten, shells, leather, 3-finger	.43
1 pr	Mitten, inserts, wool & nylon, 3-finger	.22
1 pr	Socks, wool, cushion sole	.19
1 pr	Boots, combat, rubber, black (insulated)	5.50

Quantity	Nomenclature	Weight	
1 ea	Muffler, wool	.38 lbs	
1 ea	Overwhite set, c/o: Parka	1.94	
	Trousers	1.00	
	Mitten, shell	.31	
	Clothing total		23.97 lbs

(b) Items Carried in Pockets:

Quantity	Nomenclature	Weight	
1 ea	Chapstick	.10 lbs	
1 ea	Thong, emergency	.12	
1 pr	Glasses, sun, w/case	.30	
1 ea	Box, match, waterproof	.10	
1 ea	Pocketknife (general purpose)	.40	
1 ea	Tablet, heat	.10	
1 ea	Spoon (or fork)	.10	
1 pr	Socks, wool, cushion sole (extras)	.19	
1 ea	Ski wax, blue and red	.25	
1 ea	Packet, first aid	.31	
1 ea	Canteen (⅔ full) w/cup and cover	3.00	
1 ea	C-Rations for one meal (⅓ Ration, individual, combat)	1.75	
1 ea	Necessities—handkerchief, comb, cigarettes, candy, etc.	.50	
	Pockets total		7.22 lbs
	(c) Skis, cross-country, w/bindings and ski poles		9.50
	Existence Load Total		40.69 lbs

(2) Battle Load:

(a) Weapon, Ammunition and Tools:

Quantity	Nomenclature	Weight	
1 ea	Rifle, cal .30, M-1	9.50 lbs	
1 ea	Knife-bayonet w/scabbard	.90	
2 ea	Bandoleers-cal .30 ammunition	6.75	
2 ea	Grenades, hand, frag (M-26)	2.00	
1 ea	Entrenching tool w/carrier	3.56	
	Weapon, ammunition, tools total		22.71 lbs
	(b) Mask, Protective, w/Canister and Case		3.30
	Battle Load Total		26.01 lbs

(3) Combat Load:

Quantity	Nomenclature	Weight	
	Existence Load	40.69 lbs	
	Battle Load	26.01	
	Combat Load Total		*66.70 lbs

b. **Full Field Load** worn, carried, and transported when weather is expected to change from moderate cold to severe cold (b, fig. 1).

(1) Combat Load (same as a(3) above) ____ *66.70 lbs

(2) Protection and Comfort Load:

(a) In rucksack:

Quantity	Nomenclature	Weight
1 ea	Rucksack, w/frame and white cover	5.00 lbs
1 ea	Parka, cotton-nylon, OG, 107	3.31
1 ea	Liner, parka, mohair frieze, OG, 107	2.94

* Worn and carried by the individual.

Quantity	Nomenclature	Weight
1 ea	Trousers, cotton, OG, QM shade 107, M-1951	1.12 lbs
1 ea	Liner, trousers, arctic, mohair frieze, natural	1.94
1 pr	Boots, mukluk, man's, cotton duck, natural, w/suede leather sole	1.00
2 pr	Insoles, felt	.50
2 pr	Socks, wool, natural (ski)	.75
1 pr	Socks, wool, white (felt)	.87
1 ea	Mitten set, arctic, cotton, Oxford, gauntlet, OD, type 1	1.44
1 pr	Mitten, inserts, wool and nylon, 3-finger	.22
1 ea	Toilet articles, w/towel, turkish	2.14
	Rucksack total	**21.23 lbs

(b) In Duffle Bag:

Quantity	Nomenclature	Weight
1 ea	Undershirt, man's (winter)	.94 lbs
1 ea	Drawers, men's (winter)	.94
3 pr	Socks, wool, cushion sole	.56
2 ea	Handkerchiefs	.12
1 ea	Towel, turkish	.64
1 ea	Belt, cartridge	1.50
1 ea	Helmet, steel, w/liner	3.00
1 ea	Bag, sleeping, arctic, c/o:	
	Bag, mountain (inner bag)	5.41
	Bag, arctic (outer bag)	7.06
	Case, water repellent	2.25
1 ea	Mattress, pneumatic	3.00
1 ea	Poncho, lightweight, OD w/hood	2.00
1 ea	Bag, waterproof, clothing	1.00
1 ea	Bag, duffle	2.50
	Duffle Bag Total	***30.92 lbs
	Protection and Comfort Load	52.15 lbs

(3) Full Field Load:

	Combat Load	66.70
	Protection and Comfort Load	52.15
	Full Field Total	118.85 lbs

c. **Combat Load worn and carried in extremely cold weather** (c, fig. 1).

(1) Existence Load:

(a) Clothing:

Quantity	Nomenclature	Weight
1 ea	Undershirt, man's (winter)	.94 lbs
1 ea	Drawers, men's (winter)	.94
1 ea	Trousers, men's, cotton, wind resistant, sateen, 9 oz wt, OG, QM shade 107, water repellent treated	2.25
1 ea	Liner, trousers, arctic, field	1.62
1 ea	Shirt, wool, OG, 108	1.50

** Normally carried on unit transportation. May be man-packed in an emergency.
*** Normally carried on unit transportation.

Quantity	Nomenclature	Weight
1 ea	Coat, single breasted (field jacket)	3.25 lbs
1 ea	Liner, mohair frieze (field jacket liner)	2.06
1 ea	Hood, winter, OG, w/fur ruff	.75
1 ea	Cap, pile	.44
1 ea	Suspenders, trousers	.25
1 ea	Mitten Set, arctic, cotton, oxford gauntlet, OD, type 1	1.44
1 pr	Mitten, inserts, wool and nylon, 3-finger	.22
1 ea	Trousers, cotton, OG, QM shade 107, M-1951	1.12
1 ea	Liner, trousers, arctic, mohair, frieze, natural	1.94
1 ea	Parka, cotton-nylon, OG, 107	3.31
1 ea	Liner, parka, mohair, frieze, OG, 107	2.94
1 pr	Boots, mukluk, man's, cotton duck, natural, w/suede leather sole	1.00
1 pr	Socks, wool, cushion sole	.19
2 pr	Socks, wool, natural (ski)	.75
1 pr	Socks, wool, white (felt)	.87
2 pr	Insoles, felt	.50
1 ea	Muffler, wool	.38
1 ea	Overwhite Set, c/o: Parka	1.94
	Trousers	1.00
	Mitten, shell	.31

        Clothing total _____ 31.91 lbs
   (b) Items carried in pockets (same as a(1)(b) above) _ 7.22
   (c) Skis, cross-country, w/bindings and ski poles _____ 9.50
        Existence Load Total _____ 48.63 lbs
 (2) Battle Load (same as a(2) above) _____ 26.01 lbs
 (3) Combat Load:
        Existence Load _____ 48.63 lbs
        Battle Load _____ 26.01
        Combat Load Total _____ *74.64 lbs

d. Full Field Load worn, carried and transported in extremely cold weather (d, fig. 1).

   (1) Combat Load (same as c(3) above) _____ *74.64 lbs
   (2) Protection and Comfort Load:
     (a) In rucksack:

Quantity	Nomenclature	Weight
1 ea	Rucksack, w/frame and white cover	5.00 lbs
1 ea	Bag, sleeping, arctic, c/o:	
	Bag, mountain (inner bag)	5.41
	Bag, arctic (outer bag)	7.06
	Case, water repellent	2.25
1 ea	Mattress, pneumatic	3.00
1 ea	Poncho, lightweight, OD, w/hood	2.00
1 ea	Toilet articles, w/towel, turkish	2.14

        Rucksack total _____ **26.86 lbs

---

* Worn and carried by the individual.
** Normally carried on unit transportation. May be man-packed in an emergency.

(b) In Duffle Bag:

Quantity	Nomenclature	Weight
1 ea	Undershirt, man's (winter)	.94 lbs
1 ea	Drawers men's (winter)	.94
1 pr	Mittens, shell, leather, 3-finger	.43
1 pr	Mittens, inserts, wool and nylon, 3-finger	.22
1 pr	Boots, combat, rubber, black (insulated)	5.50
3 pr	Socks, wool, cushion sole	.56
2 ea	Handkerchiefs	.12
1 ea	Towel, turkish	.64
1 ea	Belt, cartridge	1.50
1 ea	Helmet, steel, w/liner	3.00
1 ea	Bag, waterproof, clothing	1.00
1 ea	Bag, duffle	2.50

Duffle Bag Total ---***17.35 lbs
Rucksack total (from p. 280) --- 26.86 lbs
Protection and Comfort Load Total --- 44.21 lbs

(3) Full Field Load:

Combat Load --- 74.64
Protection and Comfort Load --- 44.21
Full Field Load Total --- 118.85 lbs

*** Normally carried on unit transportation.

GLOSSARY

Active layer (annually thawed layer)—Layer of ground that thaws in the summer and freezes again in the winter (equivalent to seasonally frozen ground).

Breakup—Spring melting of snow and ice, the water seeping into the ground and thawing the frozen material, which stabilizes the roadbeds, groundworks, etc. Caused by sharply rising temperatures. The ice cover of lakes, streams, and swamps becomes weak. The breakup season causes the most difficult problems in transportation and tactical operations.

Cold Injury—An inclusive term applied to injuries resulting from cold. Injuries arising from exposure to temperatures below freezing are called frostbite; injuries resulting from low temperatures, but above freezing are termed trenchfoot, immersion foot, or shelter foot, depending on the conditions of exposure.

Cornice—An overhanging formation of snow, usually on a ridge or at the top of a gully on a mountainside.

Crack—A break in the ice that can be jumped across.

Cyclonic Storms—A storm system of winds, often violent, with abundant precipitation and a usual diameter of 50 to 900 miles. It is characterized by winds rotating about a calm center of low atmospheric pressure, often at speeds as high as 90 to 130 mph. These storms are called hurricanes in the West Indies. The winds rotate clockwise in the Southern Hemisphere and counterclockwise in the Northern Hemisphere.

Disposal Bags—Heavy waterproof bags into which personnel defecate, due to the impracticability of preparing pit latrines in swampy or frozen ground.

Edging—To place or hold a ski at a different angle than that of the supporting snow.

Fall Line—The imaginary line running directly down a slope in relation to the skier. The line down which a ball of snow would roll.

Fast Ice—All types of ice, broken or unbroken, attached to the shore, beached, stranded, or attached to the bottom in shoal water.

Freezeup—The period in the autumm just before the ground is frozen for the winter. Rain and alternate periods of freezing and thawing result in a soggy terrain which presents difficult problems in transportation and tactical operations.

Frost Boil—Accumulation of excess water and mud in subsurface materials during spring thawing. It usually weakens the surface and may break through, causing a quagmire.

Frost Mound—A localized uplift of land surface caused by frost heaving or by ground water pressure. Also called earth mound, earth hummock, pals, pingo, or pingok.

Frost Table—More or less irregular surface that represents the depth of penetration of the winter frost in the seasonal frozen ground. It may or may not coincide with the permafrost table.

Fuel Tablets—Concentrated chemical fuel dispensed in tablet form for quick heating of frozen rations.

Fur Ruff—Extension of parka hood made of wolverine or similar fur, protecting the face against winds.

Greenwich Hour Angle—The angle at the pole between the meridian of Greenwich and the hour circle of any celestial body. It is always measured along the celestial equator westward from the meridian of Greenwich.

Icecrete—A mixture of sand, gravel, and water poured into forms and frozen. The process is much the same as making concrete except that ice (instead of cement) forms the bonding material.

Icing—A mass of surface ice formed by successive freezing of sheets of water that may seep from the ground, from a river, or from a spring. When the ice is thick and localized it is called an icing mound, and when it survives the summer it is called *taryn*.

Layer Principle—A clothing principle by which insulation is attained by trapping dead air in the spaces between successive layers of clothing.

Packboard—A lightweight, rectangularly shaped frame, fitted with shoulder straps and bindings. It facilitates carrying loads on men's backs by proper distribution of weight.

Poling—A pushing movement of arms and body with the ski poles against the snow to increase momentum in the slide. Single poling is referred to when each pole is used alternately to obtain this propulsion. Double poling is the use of both poles at the same time.

Sky Map—The mirroring of land, water, snow, in the clouds. A sky map approaches perfection as the clouds on an overcast day approach uniformity.

Swede Saw—A type of buck saw with a frame made of curved metal tube which provides tension to the blade and serves as a handle.

Tumpline—A kind of sling formed by a strap slung over the forehead or chest and used by one carrying a pack on his back.

Umiak—Eskimo boat about 30 feet long and 8 feet wide. Usually constructed of wooden framework over which skins are stretched. It is propelled with broad paddles.

Water Sky—Dark patches or streaks on the clouds due to the reflection of leads and polyneas, or a uniform black due to an open sea in the vicinity of large areas of ice- or snow-covered land. Details of the arrangement of the ice can be seen clearly when low stratus clouds are present. (See Sky Map).

Whiteout—An overcast sky and snow-covered terrain combine to create a condition of visibility which makes recognition of irregularities in terrain very difficult. Fog will sometimes create a similar condition.

Williwa—A sudden violent gust of cold land air, common along mountainous coasts of high latitudes.

INDEX

	Paragraphs	Pages
Action When Lost:		
Conduct of Individual	142	179
Ground-to-air signals (Code)	App. III	252
Planning of movement	140	178
Within known locality	141	178
Aerial:		
Photographs	169	215
Reconnaissance	168	215
Aircraft:		
Fixed-wing	148	190
Rotary-wing	148	190
Airfields and Airstrips	149	191
Ammunition:		
General	154	194
Storage:		
Bivouacs	84	92
Positions	154	194
Avalanche	76, 179	85, 228
Bathing. (*See* Hygiene & First Aid.)		
Battle load. (*See* Individual loads.)		
Bivouac:		
Breaking	78, 86	87, 96
Establishment	77	86
Routine	79–85	88
Security	77	86
Sites	72	82
"Blacking Out". (*See* Dehydration.)		
Body:		
Cleanliness	60	72
Heat	8, 9	11, 12
Parasites	69	80
Protection	8	11
Breakup Season. (*See* Mobility.)		
"Buddy" System:	8, 61, 151	11, 73, 192
Camouflage:		
Bivouacs	72, 173	82, 218
Clothing	16, 171	27, 217
Equipment	172	218
General considerations	164–174	210
Smoke	83	92
Tracks	73, 77, 135, 169	83, 86, 164, 215
Vehicles	174	219

	Paragraphs	Pages
Carbon Monoxide Poisoning. (*See* Hygiene and First Aid.)		
Chilling. (*See* Hygiene & First Aid.)		
Christmas Tree. (*See* Drying.)		
Clothing (Individual):		
Adjustment	3, 8, 9	5, 11, 12
Basis of issue	3	5
Camouflage	16	27
Components	12–14	16
Layer principle	7, 12	11, 16
Load	4, App. VI	5, 277
Maintenance	17	28
Principles of design	9	12
Proper use	9–11, 134	12, 163
Purpose	8	11
Weights. (*See* Weights.)		
Cold Injury	8, 61, 134	11, 73, 163
Cold Weather:		
Conditions	7	11
Operations	146, 147	186, 187
Combat:		
Characteristics	150, 151	192
Defensive techniques	157–162, App. V	195, 269
Load. (*See* Individual load.)		
Offensive techniques	157–164	195
Concealment:		
Bivouacs	72–76	82
General considerations	165	214
Individuals and groups	171–173	217
Ski trails and tracks	73, 77, 135, 169	83, 86, 164, 215
Vapor clouds (ice fog)	166	214
Condensation	78	87
Constipation	65	78
Construction of:		
Bough bed	81	90
Field latrine	77	86
Improvised shelters	37–40	47
Positions	156	195
Smudge	85	94
Storage	84	92
Trash and garbage disposal	77	86
Water point	82	91
Weapon and ski stand (rack)	77, 84	86, 92
Cover	156, 159, 164	195, 200, 210
Deception	175	221
Dehydration	59	70
Use of liquids	44	56

	Paragraphs	Pages
Digging of:		
Ditches for drainage	74	83
Frozen ground	161	202
Positions	157, 161	195, 202
Tents into the snow	75, 161	84, 202
Weapons emplacements	157	195
Discipline:		
Dehydration	59	70
Garbage disposal	82	91
Heat	42, 79	52, 88
Hood	13	19
Personal hygiene	60	72
Shelter	78	87
Track	73, 77, 134, 169	83, 86, 163, 215
Water purification	55	66
Drying:		
Clothing	10, 80	14, 89
Use of "Christmas Tree"	80	89
Environmental Effects on:		
Clothing	6	10
Instruments	152	193
Mobility	132	161
Operations	150, 151	192
Shells and grenades	160	202
Weapons	152, App. V	193, 269
Equipment:		
Availability	134	163
Effects on movement	128	153
Group	25–31, 36, 48, 161	35, 46, 58, 202
Individual	12–22, 81–85, App. VI	16, 90, 277
Load	4, App. VI	5, 277
Oversnow	14, 88, 94–97, 122–129	24, 98, 103, 145
Evacuation of:		
Battle casualties	68	80
Frostbite casualties	61	73
Existence Load. (*See* Individual Load.)		
Explosives	161	202
Field Sanitation	86	96
Fire:		
Base	43, 79	54, 88
Environmental effects	160	202
Guard	42	52
Hazard	78, 79	87, 88
Observation and control	156	195
Prevention	79	88
Small arms	157	195
Starting and maintaining	43, 79	54, 88

	Paragraphs	Pages
Fire—Continued		
Types	43	54
Wood	79, 83	88, 92
Food:		
Balanced meals	44	56
Natural resources	50–54	61
Ration types	45–47	57
Footgear:		
Principles of wear	14	24
Types for cold weather	14	24
Forests	132	161
Freezing	7, 61	11, 73
Freeze-up season. (*See* Mobility.)		
Waterways and swamps	164	210
Frostbite	8, 61	11, 73
Fuel:		
Economy	32, 79	43, 88
Firewoods	79, 83	88, 92
Storage in bivouac	84	92
Full Field Load. (*See* Individual load.)		
Ground-to-Air Emergency Code	App. III	252
Handgear	12, 13	16, 19
Headgear	12, 13	16, 19
Heat:		
Balance	8	11
Loss	8	11
Heating of:		
Lean-to	37	47
Shelter	79	88
Snow cave	42	52
Tents	30, 31	41, 43
Hoarfrost	78	87
Hygiene and First Aid:		
Application of tourniquet	67	79
Bathing facilities	60, 82	72, 91
Body parasites	69	80
Carbon monoxide poisoning	66	79
General Considerations	61–67	73
Insects	70, 85	81, 94
Waste disposal. (*See* Field Sanitation.)		
Ice:		
Conditions	135	164
Effects on:		
Mines	176	223
Shells and grenades	160	202
Weapons	153, App. V	193, 269
"Icecrete"	157	195
Ice Fog	156, 166, App. V	195, 214, 269
Load bearing capacity	135	164

		Paragraphs	Pages
Ice—Continued			
Effects on—Continued			
Melting		56, 57, 132	69, 161
Reinforcements		143	179
Strength for cover		159	200
Individual Load:			
Types		4, App. VI	5, 277
Weights		5, App. VI	7, 277
Initiative		151	192
Insect:			
Bar, nylon netting		36	46
Control		74, 85	83, 94
Types		70	81
Preventive measures		70	81
Instruments:			
Care and maintenance		155	194
Environmental effects		152, 155	193, 194
Insulation:			
Clothing		9	12
Ground		81	90
Snow cave		42	52
Land Navigation		135–140	164
Layer:			
Components of:			
Cold-dry weather uniform		13	19
Cold-wet weather uniform		12	16
Intermediate		12	16
Outer		12	16
Principle		7, 9	11, 12
Leadership		182–185	232
Lean-to. (*See* Shelters.)			
Living off country		51–54	61
Loads:			
Distribution		134	163
Individual		4, App. VI	5, 277
Weights		5, App. VI	7, 277
Maintenance of:			
Clothing and equipment		17	28
Skis and snowshoes		97, 123	114, 147
Messing		48, 49	58, 60
Mines		176–178	223
Mobility:			
Influence of:			
Break-up and freeze-up		132, 152	161, 193
Terrain		132, 143, 144, 163	161, 179, 181, 206
Weather		132, 148	161, 190
Individual		88, 131	98, 161
Rates of movement		88	98
Techniques		88, 130	98, 157
Oversnow equipment		88	98
Moisture		8	11

	Paragraphs	Pages
Movement:		
Air	148	190
Barriers	131, 132, 163	161, 206
Cross-country	88, 92, 122, 133, 134	98, 102, 145, 163
Effects of individual equipment	128	153
Environmental effects	91, 131	100, 161
Foot	134	163
Mechanized aids	143–148	179
Skijoring	130	157
Rates of movement	88	98
Requirement for trailbreaking. (*See* Trailbreaking.)		
Observation:		
Fire	156	195
Obstacles:		
Artificial	176–180	223
Crossing technique	127	152
Improving after snowfall	181	231
Natural	179	228
Swamps	164	210
Operations:		
Characteristics	23	34
Composition of units	24	35
Rescue	App. III	252
Overheating	10, 12, 13	14, 16, 19
Penetration:		
Ice	159	200
Icecrete	159	200
Snow (various types)	159	200
Timber	157	195
Principles:		
Application of basic	10	14
Clothing design	7, 9	11, 12
Feeding	44	56
Keeping warm	10	14
Layer	7, 9	11, 12
Leadership	183	233
Wearing footgear	14	24
Protection:		
Avalanches	179	228
Body	8	11
Protection and comfort load. (*See* Individual load.)		
Small arms fire	157	195
Quartering	135	164
Rations. (*See* Food.)		
Rats and mice. (*See* Field sanitation.)		
References	App. I	237
Responsibilities of Unit leader	6, 77	10, 86
Rescue	App. III	252

	Paragraphs	Pages
Roads:		
Lack	131	161
Winter	132, 143	161, 179
Route:		
Influence of:		
Seasonal changes	132	161
Terrain	132	161
Weather	132	161
Obstacles	135	164
Reconnaissance	134, 137	163, 175
Selection	135	164
Water	131, 132, 135	161, 164
Safety Measures:		
Against:		
Asphyxiation	42	52
Avalanches	179	228
Chilling	134	163
Dehydration	59	70
Fire	42, 71, 79	52, 82, 88
Insects	85	94
"Buddy" system	151	192
Prevention of accidents	30, 71	41, 82
Health hazards	61–66	73
Search	App. III	252
Security of:		
Bivouacs	135	164
Marches	135	164
Shelters:		
Construction of improvised	35–40	45
Living in	78	87
Need for	26	36
Shock. (*See* Hygiene & First Aid.)		
Ski:		
Binding	14, 94	24, 103
Care and storage	97	114
Drill	App. IV	257
Equipment	94, 158	103, 197
Pole	94	103
Preparation	95, 96	107, 109
Racks (stands)	84	92
Skijoring	130, 135	157, 164
Stacking	84, App. IV	92, 257
Waxing	96	109
Skiing:		
Advantages and disadvantages	92	102
Consideration of mobility	88	98
Preparation of equipment	95, 96	107, 109
Techniques	98–126	119
Skijoring:		
Effect on mobility	130	157
Techniques	130	157

	Paragraphs	Pages
Sled:		
Handling and placement in bivouac	84	92
Harness, sled, single trace	146	186
Pulling techniques	129	154
Types	146, 147	186, 187
Use of triple-track	135	164
Sleeping:		
Arrangements in shelter	81	90
Bough bed	81	90
Smudge. (*See* Insect.)		
Snow:		
Blindness. (*See* Hygiene & First Aid.)		
Characteristics	90	100
Classification	95	107
Composition	89	99
Conditions	135	164
Drifts	75, 164	84, 210
Effects on:		
Mines	176	223
Shells and grenades	160	202
Weapons	153, App. V	193, 269
Fences	161	202
Melting	56, 57, 132	69, 161
Reinforcements	143	179
Shelters	40–42	50
Shoes	88	98
Strength for cover	158	197
Snowshoeing:		
Need for mobility	88, 122	98, 145
Techniques	124	147
Soil condition:		
Effects on shells and grenades	160	202
Sounds	167	215
Storage	84	92
Stove:		
Operating procedures	30, 79	41, 88
Types:		
Tent, M-1941	31	43
Yukon, M-1950	30	41
Summer:		
Effects on:		
Fire (all weapons)	159	200
Mobility	132	161
Sunburn. (*See* Hygiene & First Aid).		
Sunlight:		
Effects on snow composition	89	99
Tanks	145	186
Team System	151	192
Temperature:		
Effects on snow composition	89	99
Tentage:		
Description	26, 27	36
Heating	30, 31, 79	41, 88

	Paragraphs	Pages
Tentage—Continued		
Lighting	33	45
Pitching and striking	28, 75, 161	38, 84, 202
Tools	34	45
Use in combat	161	202
Ventilation	29	41
Terrain:		
Barren tundra	75	84
Effects on mobility	91, 132	100, 161
Forests	73	83
Hazards	71	82
Marshy ground	74, 159, 160	83, 200, 202
Mountains	76	85
Suitability for trailbreaking	135	164
Thawing	7	11
Tools:		
Placing inside of shelter	78	87
Requirements for:		
Digging in frozen ground	161	202
Small unit	34	45
Trailbreaking	134–136	163
Effects of:		
Terrain	91, 132, 135	100, 161, 164
Weather	135	164
Trails:		
Camouflage and concealment	73, 77, 135, 169	83, 86, 164, 215
Types	135	164
Training:		
Objectives for military skiing	93	102
Snowshoeing	125	149
Requirement for cold weather operations.	150	192
Trench Foot. (*See* Hygiene & First Aid.)		
Types of Loads. (*See* Individual Load.)		
Uniform:		
Cold-dry weather	13	19
Cold-wet weather	12	16
Vehicles:		
Development of vapor clouds	166	214
Maintenance	App. II	240
Operational procedures	App. II	240
Placing in bivouacs	84	92
Types	143–145	179
Ventilation:		
Clothing	9	12
Tentage	29	41

	Paragraphs	Pages
Visibility:		
Effects on:		
Land navigation	138	176
Operations	168	215
Trailbreaking	135	164
Ice Fog	156, App. V	195, 269
Water:		
Availability in cold regions	55	66
Consumption	44	56
Dehydration. (*See* Dehydration.)		
Points	55, 82	66, 91
Purification	55	66
Sources	55, 56	66, 69
Transportation	55	66
Ways	131, 132, 135	161, 164
Weapons:		
Care, cleaning and maintenance	153	193
Environmental effects	152, 156, 166, App. V	193, 195, 214, 269
Handling in combat	163	206
Hanging	84	92
Positions in summer and winter	157	195
Stand (rack)	77, 78, 84	86, 87, 92
Supports used in snow	158	197
Weather:		
Blizzards	131	161
Blowing snow	156	195
Effects on:		
Infantry weapons	App. V	269
Trailbreaking	135	164
Fog	156	195
Ice fog phenomenon	166	214
Limitations on air operations	148	190
Seasonal changes	132	161
Sudden changes	134	163
Weights:		
Individual clothing and equipment	4–6, App. VI	5, 277
Individual loads	4, 5, App. VI	5, 7, 277
Battle	5, App. VI	7, 277
Combat	5, App. VI	7, 277
Existence	5, App. VI	7, 277
Full Field	5, App. VI	7, 277
Protection and comfort	5, App. VI	7, 277
Wind:		
Effects on:		
Bivouacs	75, 76	84, 85
Combat	156, 164	195, 210
Ice	157	195
Snow composition	89	99

	Paragraphs	Pages
Wind—Continued		
"Falls"	127	152
Winter:		
Effects on mobility	132	161

[AG 353 (18 Dec 58)]

By Order of *Wilber M. Brucker*, Secretary of the Army:

MAXWELL D. TAYLOR,
General, United States Army,
Chief of Staff.

Official:
R. V. LEE,
Major General, United States Army,
The Adjutant General.

Distribution:
Active Army:
    DCSPER (1)
    ACSI (1)
    DCSOPS (1)
    DCSLOG (1)
    Tech Stf, DA (1)
    Tech Stf Bd (1)
    USCONARC (10)
    OS Maj Comd (5) except
      USARAL (400)
    Log Comd (1)
    MDW (1)
    Armies (5)
    Corps (5)
    Div (5) (Ea CC) (2)
    Regt/Gp/bg (2)
    Bn (1)
    Co (1)
    USMA (2)
    USCGSC (800)
    Br Svc Sch (10) except
      USAIS (44), USAARMS (25), USASCS (20), USA CmlC SCH (15), USA QM Sch (15)
    Mil Msn (1)
    Mil Dist (1)
    USA Corps (Res) (1)
    Sector Comd (Res) (1)
    MAAG (1)

NG: State AG (3); Div (3) (Ea CC) (2); Regt/Gp/bg (2)

USAR: units—same as Active Army except allowance is one copy to each unit.

For explanation of abbreviations used, see AR 320-50.

www.ingramcontent.com/pod-product-compliance
Lightning Source LLC
Chambersburg PA
CBHW032359100526
44587CB00011BA/735